Mathematical Thinking and Learning

An International Journal

Volume 5, Numbers 2&3, 2003

Special Issue:
Models and Modeling Perspectives

Guest Editor
Richard Lesh

Contents

NEW YORK AND LONDON

Journal Information

Subscriptions: *Mathematical Thinking and Learning: An International Journal* is published quarterly by Lawrence Erlbaum Associates, Inc., 10 Industrial Avenue, Mahwah, NJ 07430–2262. Subscriptions for the 2003 volume are available only on a calendar-year basis.

Individual rates: **Print *Plus* Online:** $55.00 in USA, $85 outside USA. Institutional rates: **Print-Only:** $295.00 in USA, $325.00 outside USA. **Online-Only:** $295.00 in USA and outside USA. **Print *Plus* Online:** $325.00 in USA, $355.00 outside USA. Visit LEA's Web site at http://www.erlbaum.com to view a free sample.

Order subscriptions through the Journal Subscription Department, Lawrence Erlbaum Associates, Inc., 10 Industrial Avenue, Mahwah, NJ 07430–2262.

Claims: Claims for missing copies cannot be honored beyond 4 months after mailing date. Duplicate copies cannot be sent to replace issues not delivered due to failure to notify publisher of change of address.

Change of Address: Send address changes to the Journal Subscription Department, Lawrence Erlbaum Associates, Inc., 10 Industrial Avenue, Mahwah, NJ 07430–2262.

Permissions: Special requests for permission should be sent to the Permissions Department, Lawrence Erlbaum Associates, Inc., 10 Industrial Avenue, Mahwah, NJ 07430–2262.

Abstracts/Indexes: This journal is abstracted or indexed in *PsycINFO/Psychological Abstracts; Zentralblatt für Mathematik/Mathematics Abstracts; ERIC Clearinghouse for Science, Mathematics, and Environmental Education; Current Index to Journals in Education; EBSCOhost Products; Education Index; Education Abstracts; and Australian Education Index.*

Microform Copies: Microform copies of this journal are available through ProQuest Information and Learning, P.O. Box 1346, Ann Arbor, MI 48106–1346. For more information, call 1–800–521–0600, ext. 2888.

First published by Lawrence Erlbaum Associates, Inc., Publishers
10 Industrial Avenue
Mahwah, NJ 07430

ISSN 1098–6065

This edition published 2013 by Routledge

Routledge
711 Third Avenue
New York
NY 10017

Routledge
2 Park Square, Milton Park
Abingdon
Oxon, OX14 4RN

Routledge is an imprint of the Taylor & Francis Group, an informa business

MATHEMATICAL THINKING AND LEARNING, 5(2&3), 109–129

Models and Modeling Perspectives on the Development of Students and Teachers

Richard Lesh
School of Education
Purdue University

Richard Lehrer
Department of Teaching and Learning
Peabody College, Vanderbilt University

This special issue of *Mathematical Thinking and Learning* describes *models and modeling perspectives* toward mathematics problem solving, learning, and teaching (Lesh & Doerr, 2003). The term "*models*" here refers to purposeful mathematical descriptions of situations, embedded within particular systems of practice that feature an epistemology of model fit and revision. That is, "modeling" is a process of developing representational descriptions for specific purposes in specific situations. It usually involves a series iterative testing and revision cycles in which competing interpretations are gradually sorted out or integrated or both—and in which promising trial descriptions and explanations are gradually revised, refined, or rejected. The latter emphasis on the "fitness" of models is critical because it suggests that models are inherently provisional, and it emphasizes that they are developed for specific purposes in specific situations—even though they may endure for longer periods of time, and even though they generally are intended to be sharable and reuseable in a variety of structurally similar situations.

The distinction between model and world is not merely a matter of identifying the right symbol-referent matches; rather, it depends intimately on the accumulation of experience and its symbolic representations over time. Models bootstrap

Requests for reprints should be sent to Richard Lesh, School of Education, BRNG 1440 Room 6130, Purdue University, West Lafayette, IN 47907–1440. E-mail: rlesh@purdue.edu the world and the world "pushes back" toward revision of one's models. The result is a "mangle of practice" (Pickering, 1995) in which argument, mathematics, and nature are comingled.

In this special issue, we are concerned not only with mature forms of models and modeling in communities of scientists and mathematicians, but also with the need to initiate students in these forms of thought. When attention is turned toward students, it is important to account for factors that lead students to recognize the need for a given type of model—and to account for students' developing models and for their developing participation in practices that involve the invention and revision of models. These practices are constituted to negotiate the interchange between the world and its mathematical counterparts.

The four contributions to this special issue suggest a variety of ways that students (children through adults) can be introduced to highly productive forms of modeling practices. Collectively, the contributions illustrate how modeling activities often lead to remarkable mathematical achievements by students formerly judged to be too young or too lacking in ability for such sophisticated and powerful forms of mathematical thinking. They also illustrate how modeling activities often create productive interdisciplinary niches for mathematical thinking, learning, and problem solving that involve simulations of similar situations that occur when mathematics is useful beyond school. Petrosino, Lehrer, and Schauble lead off by describing how elementary school children (Grade 4 in the United States) developed a conceptual system about the notion of distribution as a model for structuring errors of measure. These children then considered distributions of results obtained via experiment in light of their emerging understanding of the structure of error as distributed. Lesh and Harel focus on the developing knowledge of middle school or high school students—and on their abilities to engage in activities that involve a variety of different types of proportional reasoning. Schorr and Clark shift attention toward the developing knowledge of teachers as exhibited in the ways that they make sense of their students' work in model eliciting activities (Lesh, Hoover, Hole, Kelly, & Post, 2000). Finally, Doerr, Lesh, Carmona, and Hjalmarson provide a brief overview of several of the most significant ways that models and modeling perspectives imply the need to reconsider a number of currently popular "constructivist" views about: (a) the nature of children's developing mathematical knowledge; (b) the nature of real life situations (beyond school) where elementary, but powerful mathematical constructs, are useful; (c) the nature of mathematical understandings and abilities that contribute to success in the preceding problem-solving situations; and (d) the nature of teaching and learning situations that contribute to the development of the preceding understandings and abilities. In particular, models and modeling perspectives highlight the inherent reflexivity of collective and individual thought by naturally incorporating both social and individual perspectives on the nature of mathematical knowledge—much in the manner suggested by "pragmatists" such as Mead (1910) and Dewey (1982) nearly a century ago.

WHAT ARE DISTINCTIVE CHARACTERISTICS OF MODELS AND MODELING PERSPECTIVES?

In mathematics education, models and modeling perspectives emphasize the fact that "thinking mathematically" is about interpreting situations mathematically at least as much as it is about computing. Or, in the case of mathematics teacher education, models and modeling perspectives emphasize that expertise in teaching is reflected not only in what teachers can "do," but also what they "see" in teaching, learning, and problem-solving situations. That is, it involves not only doing things right, but also doing the right things at the right time and with the right people. For example, expertise in teaching involves the development of powerful conceptual tools for making sense of students' work. Similarly, in the case of school children, the development of elementary, but powerful mathematical models, should be considered to be among the most important goals of mathematics instruction. So, answers to the question: "What mathematical knowledge and abilities has a given student developed?" should include not only information about "What computations can they do?" but also information about "What kinds of situations can they describe mathematically?" It should include information about "What models have they developed?" as well as information about "What skills have they mastered?"

Like most contemporary theories in cognitive science, models and modeling perspectives begin with the assumption that humans interpret their experiences using internal conceptual systems (or constructs) whose functions are to select, filter, organize, and transform information, or to infer patterns and regularities beneath the surface of things (Lesh & Doerr, 2003). Yet, whereas some cognitive theories speak about these interpretive or descriptive systems as if they resided totally within the minds of students, models and modeling perspectives recognize that to have sufficient power for dealing with realistically complex problem-solving situations, relevant conceptual systems usually must be expressed using a variety of interacting media that may range from spoken language, to written symbols, to diagrams, to experience-based metaphors, to computer-based simulations (Cramer, 2003; Johnson & Lesh, 2003). In short, thinking is a form of mediated activity (Wertsch, 1985); and, because different media generally emphasize and deemphasize somewhat different aspects of the underlying conceptual systems (and the situations that they are used to describe), the meanings of these conceptual systems often are distributed across a variety of interacting media. In this sense, modeling is a form of literacy (diSessa, 1998; Gee, 1997; Olson, 1994).

FOUNDATIONS OF MODELS AND MODELING PERSPECTIVES

As Figure 1 suggests, mathematical models are conceptual systems that are: (a) expressed for some specific purpose (which John Dewey referred to as an

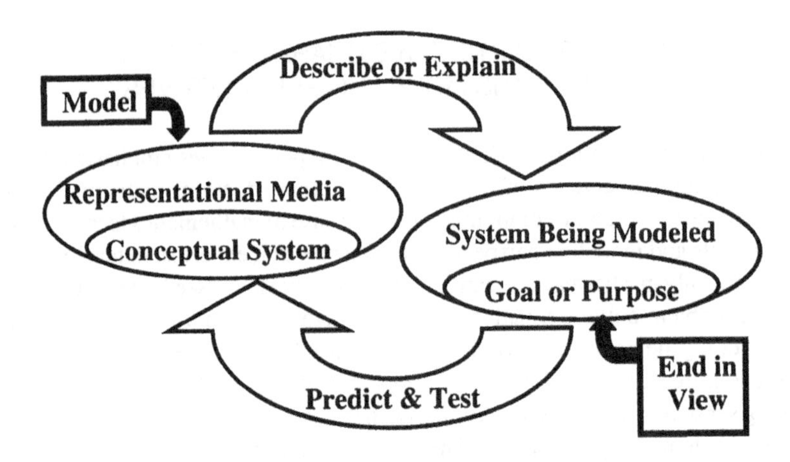

FIGURE 1 A modeling cycle.

"end-in-view"), and (b) expressed using some (and usually several) representational media. That is, mathematical models are purposeful descriptions or explanations. They focus on patterns, regularities, and other systemic characteristics of structurally significant systems. Their purposes often involve constructing, manipulating, or predicting the systems that are being modeled; and, the process of developing sufficiently useful models for a specific purpose usually involves a series of iterative testing and revision cycles (Lesh & Doerr, 2003).

As students go through a series of modeling cycles during a given modeling activity, what we expect to observe is the emergence of a series of systematically different ways of thinking about the nature of the objects, relations, operations, and patterns or regularities in the problem-solving situation. This is because mathematical models, and their underlying conceptual systems, generally are defined by specifying four components.

- The nature of their mathematical "objects" (e.g., quantities, shapes, locations),
- The nature of their mathematical relations among "objects,"
- The nature of their mathematical operations on "objects,"
- The nature of their mathematical patterns and regularities that govern the preceding objects, relations, and operations.

According to Figure 1, mathematical models involve:

- purposes,
- underlying conceptual systems,
- media in which the conceptual system is expressed.

Because models are developed for specific purposes in specific situations, they involve situated forms of learning and problem solving (Greeno, 1991). On the other hand, the need to develop models (or other conceptual tools) seldom arises unless part of the goal includes sharability (with others) and reusability (in other situations). Therefore, modeling is inherently a social enterprise, and significant forms of generalizability and transferability are involved.

AN EXAMPLE OF A MODEL-ELICITING ACTIVITY FOR MIDDLE SCHOOL MATHEMATICS

In research on models and modeling in school classrooms, the kind of model-eliciting activities that are referred to in the preceding section usually are simulations of "real life" situations in which mathematical thinking is needed for success (Lesh et al., 2000). One characteristic of such problems is that the products that students produce are not restricted to brief answers to artificially restricted questions about premathematized situations. For example, the goal often is to develop a conceptual tool that goes beyond being useful for some specific purpose in a given situation—and that also is sharable (with others) and reuseable (in other similar situations). What's problematic about most model-eliciting activities is that students must make symbolic descriptions of meaningful situations. Consequently, it is not surprising that model-eliciting activities tend to emphasize quite different understandings and abilities than those that are emphasized on traditional textbook word problems—where students must make sense of symbolically described situations.

The Paper Airplane Problem that follows is a middle school version of a "case study" that we first saw being used in Purdue's graduate program for aeronautical engineering—where students used a wind tunnel to develop a useful operational definition for the concept of "drag" for various shapes of planes and wings.

> Note: Here, our commitment to pedagogical design focused on the question: "How can we design a problem that will allow middle school students to experience some of what we observed with the college engineering students?"

When seventh-grade students began to work on The Paper Airplane Problem, they read a newspaper article that described how to make a variety of different types of paper airplanes. Then, they made several of these airplanes and tested the flight characteristics using three different types of flight paths (Figure 2). For each flight path, the goal was to hit a designated target, starting from a given starting point, and traveling around some obstacle (such as a chair). Then, each team of students was given a data sheet (like the one shown in Table 1) showing results that were produced by "another group of students" in another school. That is, for each paper airplane, results were shown for each of three flight paths; and, for each

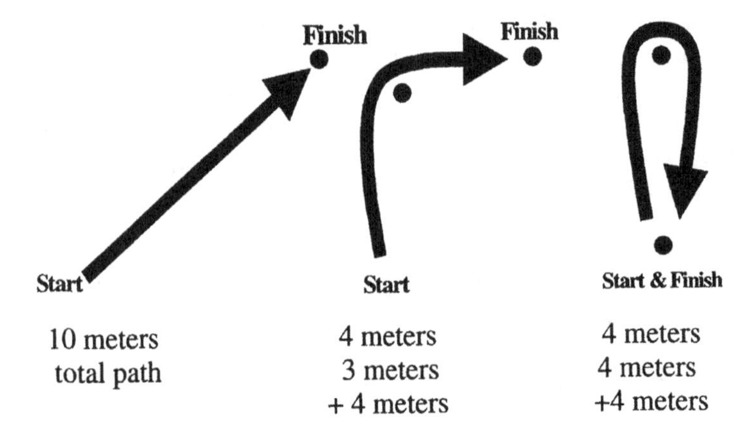

FIGURE 2 Three flight paths.

flight path, the following three measurements were recorded: (1) the total distance flown, (2) the distance from the target, and (3) the time in flight. The goal of the problem was for students to write a letter to students in another class describing how such data can be used to assess paper airplanes for following four kinds of flight characteristics: (1) best floater (i.e., going slowly for a long time), (2) most accurate, (3) best boomerang, and (4) best overall.

Notice that The Paper Airplane Problem is similar to many familiar situations that occur in "real life" situations when people need to think quantitatively about things that they can't see—or cannot measure directly. Examples include most kinds of rates (speeds, exchange rates); and, they also include indexes of things such as the "productivity" for people, products, or strategies. On our website (http://tcct.soe.purdue.edu/resources/) is a transcript showing a solution that one group of average ability seventh grade students developed for this problem. As the transcript shows, solutions to model-eliciting activities often involve sorting out and integrating concepts associated with a variety of different topic areas in mathematics and the sciences. They usually involve several modeling cycles in which trial solutions are gradually refined, revised, or rejected. They generally require students to use a great many elementary mathematical ideas and skills that are not addressed in school textbooks and tests. And, in particular, they often emphasize multimedia representational fluency, as well as a variety of mathematical abilities related to argumentation, description, and communication—as well as abilities needed to plan, monitor, and assess progress while working in teams of diverse specialists. In short, because such tasks emphasize a broader range of mathematical abilities than those emphasized tests that are easily scoreable, they often enable a broader range of students to emerge as being exceptionally capable (Lesh, 2001; Zawojewski & Bowman, 2001–2004).

TABLE 1
Paper Airplane Data Table

	Which team's plane should will the prize as being the best floater?								
	Path 1			*Path 2*			*Path 3*		
Team	*Amount of Time in Air (sec)*	*Length of Throw (m)*	*Distance From Target (m)*	*Amount of Time in Air (sec)*	*Length of Throw (m)*	*Distance From Target (m)*	*Amount of Time in Air (sec)*	*Length of Throw (m)*	*Distance From Target (m)*
Team 1	3.1	11.0	1.8	2.5	7.7	3.2	0.7	1.8	6.8
	0.1	1.5	8.7	0.9	2.9	8.6	1.2	3.7	6.7
	2.7	7.6	4.5	0.1	1.1	2.4	2.7	8.4	4.4
Team 2	3.8	10.9	1.7	3.2	9.2	4.6	2.3	8.1	6.1
	5.2	13.1	5.4	2.3	9.4	2.9	0.2	1.6	6.9
	1.7	3.4	8.1	1.1	2.7	8.8	2.1	6.9	5.2
Team 3	4.2	12.6	4.5	1.7	4.0	5.9	2.9	8.5	5.5
	5.1	14.9	6.7	3.0	10.8	3.1	2.4	7.7	8.7
	3.7	11.3	3.9	2.0	6.0	3.2	0.2	1.9	6.7
Team 4	2.3	7.3	3.25	1.3	4.9	4.4	1.4	4.9	4.9
	2.7	9.1	4.9	2.3	7.3	2.4	2.7	7.2	8.1
	0.2	1.6	9.1	1.4	3.7	4.1	0.2	1.1	7.3
Team 5	4.9	7.9	2.8	2.7	10.7	1.1	2.5	7.7	5.7
	2.5	10.8	1.7	3.3	6.3	2.5	2.1	9.8	9.8
	5.1	12.8	5.7	1.3	2.6	7.2	3.2	10.4	5.8
Team 6	0.2	1.8	8.8	1.8	3.9	4.2	0.1	1.2	8.2
	2.4	10.1	4.6	0.2	5.7	4.2	1.3	4.9	4.9
	4.7	10.3	5.4	1.6	8.5	3.4	1.8	5.5	2.7

Note: The "length of throw" indicates the straight-line distance between the starting point and where the plane landed. The "target" is the finishing point where the plane should land.

THE EMERGENCE OF MODELS AND MODELING PERSPECTIVES IN MATHEMATICS EDUCATION RESEARCH

As models and modeling perspectives have emerged as significant conceptual frameworks for systemic research on the interacting development of students, teachers, curriculum materials, and programs of instruction, three distinct, but closely related lines of inquiry have been emphasized (Lesh & Doerr, 2003). A brief description follows for each of the three. All three are represented in this special issue of Mathematical Thinking and Learning, and, all three are elaborated and extended in a new book that is titled *Beyond Constructivism: Models and Modeling Perspectives on Mathematics Problem Solving, Learning, and Teaching* (Lesh & Doerr, 2003).[1]

Research on Problem Solving Beyond School

The first line of inquiry has focused on model-eliciting problems (Lesh et al., 2000) and on two major themes: (a) identifying the mathematical understandings and abilities that are needed for success when "mathematical thinking" is needed beyond school in a technology-based age of information, and (b) identifying students who have extraordinary abilities that may not have been apparent based on past records of low performance on the narrow and shallow band of tasks emphasized in traditional textbooks and tests (Lesh, 2001). Significant findings from this research include the facts that: (a) the mathematical understandings and abilities that are emphasized on standardized tests typically represent only a remarkably narrow and shallow subset of those needed for success beyond school in a technology-based age of information (Lesh, 2001), and (b) when assessments recognize the importance of a broader range of mathematical understandings and abilities, of the type that are needed for success beyond school in a technology-based age of information, a broader range of students naturally tend to emerge as having exceptional potential (Lesh, Zawojewski, & Carmona, 2003).

---

[1]The book, *Beyond Constructivism* (Lesh & Doerr, 2003), contains chapters written by more than thirty leading researchers inside and outside the field of mathematics education. Collaborations leading to these publications grew out of three multisite research groups: (a) the Models and Modeling Working Group associated with the University of Wisconsin National Center for Improving Student Learning and Achievement in Mathematics and Science Education (NCISLA), (b) the Models and Modeling Working Group associated with the North American Group for the Psychology of Mathematics Education (PME-NA), and (c) Purdue University's Center for Twenty-first Century Conceptual Tools (TCCT).

Research on Design Principles for Productive Modeling Activities for Learning in School

The second line of inquiry is informed by the first. It focuses on the question: "If models and modeling practices are to be introduced to schools, how can this be best accomplished?" This orientation leads to investigations of design principles for productive modeling environments in school. Consequently, this line of inquiry often focuses on: (a) designing learning environments in which students develop deeper and higher-order understandings of powerful ideas in elementary mathematics and science, and (b) investigating ways to use new conceptual technologies to help make extraordinary achievements accessible to ordinary children.

This second line of inquiry often focuses on teacher–student interactions as much as on student–student interactions (Zawojewski, Lesh, & English, 2003); and, it also often focuses on the development of whole classrooms of students as learning communities that develop productive shared norms about the nature of scientific argumentation and justification (Lehrer & Schauble, 2003; McClain, 2003; vanReeuwijk & Wijers, 2002). Significant findings from this research include the fact that surprisingly young children, as well as students from highly disadvantaged backgrounds, often produce high quality results that are far more impressive than anything that would have been predicted based on results from their prior work in traditional textbooks and tests (Carlson, Larsen, & Lesh, 2002; Doerr & Lesh, 2002; Kaput & Schorr, in press, Lehrer & Schauble, 2002; Lesh, 2001; Shternberg & Yerushalmy, 2003).

Overall, this second line of inquiry complements the first. Both are attuned to the intersection of mathematics and science—as well as to the integration of diverse topics within mathematics or science. Both are aimed at helping students develop mathematical models, or sense-making systems, as a form of explanation of the natural world. Both focus on promoting conceptual development by putting students in situations where they repeatedly express-test-and-revise their own current ways of thinking about "big ideas" in mathematics or science—rather than simply being led to adopt teachers' prefabricated ways of thinking. And, both strive for deep treatments of a small number of "big ideas"—rather than being preoccupied with superficial "coverage" of a large number of lower-level facts and skills.

Research on the Nature of Teachers' Developing Knowledge and Abilities

The third line of inquiry shifts attention toward the nature of teachers' developing knowledge and abilities (Clark & Lesh, 2003; Doerr & Lesh, 2002; Lehrer & Schauble, 2000; Schorr & Lesh, 2003). Results from this research include details about significant ways that most teachers' mathematical understandings need to be enhanced to help their students expess-test-and-revise their thinking in productive directions (Schorr & Lesh, 2002).

Based on *models and modeling perspectives*, such research often involves on-the-job classroom-based professional development activities in which model-eliciting activities for students provide contexts in which teachers' teaching experiences become productive learning experiences to support teacher development (Clark Koellner & Lesh, 2003; Schorr & Lesh, 2003). Keys to the success of this approach include the following.

- Model-eliciting activities for students are activities in which students repeatedly express their current ways of thinking in forms that are visible to teachers—and to the students themselves. Therefore, as students repeatedly express, test, and revise their ways of thinking, they automatically produce auditable trails of documentation that reveal important things about the constructs and conceptual systems that they are developing. In this way, such activities are thought-revealing activities.
- Research on cognitively guided instruction has shown that one of the most effective ways to help teachers improve their teaching is to help them become familiar with their students' evolving ways of thinking about important ideas and abilities that they want their students to develop (Carpenter, Fennema, & Romberg, 1993).
- At the same time that students are developing powerful conceptual tools to make sense of model-eliciting activities, the thought-revealing nature of their responses provides opportunities for teachers to develop sharable and reusable tools that colleagues can use to observe, document, or make sense of students' work. … In research based on models and modeling perspectives, these teacher-level tools have included:

  - observation forms to gather information about the roles and processes that contribute to students' success,
  - ways of thinking sheets to identify strengths and weaknesses of products that students produce—to help teachers provide appropriate feedback and directions for improvement,
  - quality assessment guides for assessing the quality of alternative products that students produce,
  - guidelines for conducting mock job interviews based on students' portfolios of work produced during case studies for kids—and focusing abilities valued by employers in future-oriented professions.

Thus, thought-revealing activities for students often provide contexts for equally thought-revealing activities for teachers (or parents, policy makers, professors, professionals in business and industry).

In research conducted using models and modeling perspectives, the three lines of inquiry that have been described in this section often are not conducted in isolation. They are linked. For example, research on teacher development often takes place in the context of research on student development; or, research on problem solving (outside of school) often takes into account results from research on ideas and abilities being developed in school. So, it is important to investigate the interacting development of students, teachers, and programs (Lesh, 2002)—and interactions between learning and problem solving.

Several recent research publications describe multitier research designs that were explicitly created for use in research conducted using models and modeling perspectives (Kelly & Lesh, 2000; Lesh, 2001). In these multitier design studies, students may develop thought-revealing conceptual tools for use in mathematical problem-solving situations at the same time that teachers (or researchers) are developing thought-revealing conceptual tools to encourage (or make sense of) students' (or teachers') modeling activities. Yet, whereas model-eliciting activities for students are decision-making situations in which students need to use mathematical ways of thinking in their everyday lives, model-eliciting activities for teachers focus on teachers' classroom decision-making issues, and they require teachers to integrate mathematical, psychological, historical, and pedagogical ways of thinking (Lesh, 2001). ... At all three tiers of such research designs, development is encouraged because each of the interacting participants (students, teachers, researchers) repeatedly express their current ways of thinking in the form of complex artifacts that are tested and revised repeatedly. Therefore, similar design principles apply to the creation of productive knowledge development at all three levels; and, at all three levels, the development cycles that participants go through automatically produce auditable trails of documentation that reveal important information about the nature of the constructs and conceptual systems that are being developed. In other words, such design activities contribute to development while at the same time generating documentation about the nature what is being learned (Lesh & Doerr, 2002).

THEORETICAL FOUNDATIONS FOR MODELS AND MODELING PERSPECTIVES

Models and modeling perspectives build on Piaget's *structuralist* views about the holistic and constructed nature of the conceptual systems that children develop to make sense of their mathematical experiences. They build on Vygotsky's conception of thought as *mediated activity* (Wertsch, 1985). And, above all, they build on foundations established by *American Pragmatists* such as John Dewey (1982), William James (1982), and Charles Sanders Peirce (1982) who were skeptical of "grand theories" in education—and who focused on developing a "blue collar" conceptual framework for "real life" decision making by teachers and other educators (Lesh & Doerr, 2002).

One main idea that models and modeling perspectives adopt from Piaget is his emphasis on the *holistic nature of conceptual systems* that underlie people's mathematical interpretations of experiences (Beth & Piaget, 1966). Piaget emphasized that many of the most important properties of mathematical systems-as-a-whole are not derived from properties of their constituent elements. Consequently, the development of these conceptual systems must involve more than simply assembling (or constructing, or piecing together) the elements, relations, operations, and principles that they include (Lesh & Carmona, 2002).[2] Instead, emergent properties at higher-level systems evolve from (and are reflectively abstracted from) systems of interactions at more primitive/concrete/enactive/intuitive levels; and, these conceptual reorganizations occur mainly when models fail to fit the experiences they are intended to describe, explain, or predict. Then, when conceptual reorganizations are required, development occurs along a variety of dimensions—concrete–abstract, simple–complex, intuitive–formal, situated–decontextualized, specific–general—where the right side of these developmental continua are not necessarily characteristics of more advanced understanding.[3]

> Practical implications of these perspectives include the fact that, from a models and modeling perspective, a main challenge for teachers is to find ways to put students in situations where they must express, test, and revise their own current ways of thinking. The book, *Beyond Constructivism: Models and Modeling Perspectives of Mathematics Problem Solving, Learning, and Teaching* (Lesh & Doerr, 2002), gives many examples to show that, for a given "big idea" in elementary mathematics, getting students to clearly recognize the need for the underlying construct is often a large part of what it means to "understand" the construct. A large part of the meaning of the construct comes from recognizing why it is needed—and recognizing how it is related to other relevant, but logically unrelated constructs.

---

[2]All of Piaget's famous *conservation tasks* required students to make judgments about (what mathematicians refer to as) *invariance properties* under a variety of different systems of operations, relations, and/or transformations. Consequently, because a property that is invariant with respect to a system is not meaningful until students begin to use the relevant systems to interpret their experiences, tasks that assess a person's understanding of these invariance properties often are powerful tools for assessing their understanding of the relevant system (Lesh & Carmona, 2002)

[3]In a given situation, the model that is most powerful and useful is not necessarily the one that is most abstract, complex, detailed, formal, decontextualized, or general. In general, in the context of specific learning or problem solving situations, models (and accompanying conceptual systems) that are "best" are those that deal appropriately with trade-offs involving factors such as simplicity and complexity, or cost and quality.

Drawing from the Vygotskian tradition of thought as mediated activity (Wertsch, 1991), models and modeling perspectives also emphasize the roles of conceptual tools, such as those that are supported by language or notational systems, that influence the power of peoples' thinking. Because of the power of such conceptual tools, they generally have strong influences on students' mathematical thinking from both an individual endeavor and a collective enterprise. This dual relation between the collective and individual experience has a long tradition in American Pragmatism—especially in Dewey's "instrumentalist" interpretations of pragmatist perspectives (Dewey, 1998).

> Practical implications of these perspectives include the fact that, whereas Piagetians often are interpreted as being pessimistic about the possibility of significantly influencing students' levels of development of powerful constructs and conceptual systems, one important implication of Dewey's instrumentalist perspectives is reflected in the Jerome Bruner's famous claim that "Any concept can be taught to any child at any time in some intellectually respectable way" (Bruner, 1963). Although Bruner's claim clearly is an exaggeration, his central points are straightforward. First, ideas develop as they come to be expressed using increasingly powerful conceptual tools and representational media. Second, by introducing appropriate conceptual tools, surprisingly sophisticated constructs often can be made accessible to most students, including those who have been least privileged, as long as the situations are meaningful and the appropriate conceptual tools are available (Harel & Lesh, 2002, Kaput & Schorr, in press; Lehrer & Schauble, 2000).

Another main idea that models and modeling perspectives adopt from Vygotsky is the concept of *zones of proximal development* (Vygotsky, 1978; Wertsch, 1985). That is, ideas develop, and, at any given point in time, a student's level of understanding can be influenced by a variety of factors such as: (a) guidance provided by an adult or peer (Cobb & Yackel, 1998), (b) conceptual tools that may be available either by luck or because of interventions from an adult (Kaput, 1994), or (c) approaches suggested (or limited by) sociocultural norms and standards that have been developed by relevant communities—such as students and teachers in classrooms (McClain, 2002; vanReeuwijk & Wijers, 2003). Consequently, the instructional challenge is to help students extend, revise, reorganize, refine, modify, or adapt constructs (or conceptual systems) that they DO have—not simply to find or create constructs that they do not have (or that are not immediately available). However, whereas Vygotsky (1978) emphasized the influences of language on thought, models and modeling perspectives also recognize that language is only one among many culturally supplied conceptual tools that influence mathematical thinking (Cobb & McClain, 2001; Dewey, 1982). Furthermore, whereas Vygotsky focused on the internalization of external functions, models and modeling perspectives recognize that development of powerful conceptual tools occurs along a variety of dimensions (Lesh, 2002). Therefore, the notion of a zone of proximal development needs to be expanded from a 1-dimensional interval to an N-dimensional region in which a variety of paths lead to any given construct. Furthermore, students are able to make progress through these regions along a variety of possible trajectories.

> Practical implications of these perspectives include the fact that, particularly in the case of especially powerful constructs and conceptual systems, placing children in situations where they express, test, and revise their own ways of thinking is often quite different than leading them along narrow trajectories that lead to the (often superficial) adoption of our ways of thinking—especially if it is assumed that all children should develop along the same trajectory.

Another way that models and modeling perspectives extend Vygotsky's ideas about the influence of social functions on conceptual development is related to Marvin Minsky's notion of *communities of mind* (Minsky, 1987)—or William James' concept of a *pluriverse* of conceptual systems. According to Minsky and James, in nontrivial learning or problem-solving situations, students generally should be expected to have available a community of conceptual systems—each of which have the potential to be engaged to interpret relevant experiences (Zawojewski, Lesh, & English, 2003). Thus, a student's developing community of constructs is similar to a community of living, adapting, and continually evolving biological systems (Lesh & Doerr, 1998). Consequently, development is not likely to be encouraged by discouraging diversity; and, when the products that students produce include descriptions and explanations, then there always exist a variety of different types of responses—where trade-offs often must be considered among factors such as: precision versus accuracy, complexity versus timeliness, simplicity versus superficiality, power versus economy, or costs versus benefits.

> Practical implications of these perspectives include the fact that, in a community of students, as well as in a given students' community of potentially relevant constructs in a given problem-solving situation, diversity is to be encouraged—as long as it also is accompanied by selection, communication (so that innovations will spread), and conservation (so that innovations will be preserved). For example, in the transcripts that are given in digital appendixes to this special issue of *Mathematical Thinking and Learning*, it is clear that, when students make progress, they often do so by sorting out and integrating diverse (and often logically unrelated) ways of thinking. Furthermore, decisions about rejected ways of thinking often contribute as much to learning or problem solving as decisions about ways of thinking to adopt, or

refine, or revise. Consequently, a great deal of attention often is needed to ensure that ways of thinking are accepted or rejected because they are most appropriate (i.e., most useful in the given situation)—not simply because an "authority figure" says so.

CONCLUDING REMARKS ABOUT PRACTICAL ISSUES IN INSTRUCTION

Even though models and modeling perspectives can be seen as evolving naturally out of currently popular "constructivists" ways of thinking about the nature of mathematics, problem solving, learning, and teaching, they often require the development of new ways of thinking that are quite different than those traditionally adopted in mathematics education research (Lester & Kehle, 2003).

Concerning Relations Between Modeling and Problem Solving

Whereas mathematics education researchers traditionally have defined problem solving as a process of "getting from givens to goals when the solution processes are not readily available" (Schoenfeld, 1982), models and modeling perspectives emphasize that the development of models generally occurs through a series of develop-test-revise cycles–each of which involve somewhat different ways of thinking about the nature of givens, goals, and possible solution steps (Lesh & Harel, 2003). Consequently, when solution processes involve a series of modeling cycles, when early interpretations of given and goals can be expected to be naïve, and when the goal is to extend, revise, reorganize, refine, modify, or adapt constructs (or conceptual systems) that you do have rather than to function better when none are available, quite different kinds of problem-solving strategies and metacognitive processes emerge as important (Lesh, Lester, & Hjalmarson, 2003; Middleton, Lesh, & Heger, 2003; Zawojewski & Lesh, 2003).

Concerning Relations Between Applied Mathematics and Pure Mathematics

John Dewey, in particular, stressed the notion that the goal of making practice more intelligent is quite different than the goal of making intelligence more practical (Dewey, 1982). Similarly, mathematizing reality is quite different than "realizing" mathematics. This is why, models and modeling perspectives stress the importance of going beyond "teaching mathematics so as to be useful" (Freudenthal, 1973) to also help children learn to quantify, dimensionalize, coordinatize, and in other ways mathematize their experiences. On the other hand, models and modeling perspectives should not be interpreted as deemphasizing "pure" mathematical activities. Nor is it being suggested that, by taking students' immature ways of thinking seriously (so that they can be explicitly tested and revised or rejected), this implies that "anything goes" and that immature ways of thinking should be celebrated while the wisdom of sages should be ignored. The whole point of emphasizing models and modeling activities is to focus on deep treatments of a small number of "big ideas" and to increase the likelihood that students will develop powerful constructs and conceptual systems—while ensuring that they will go beyond thinking with these systems to also think about them. Indeed, in a number of recent publications, we have been explicit about a variety of ways to achieve an appropriate balance between "pure" and "applied" mathematical activities (Kelly & Lesh, 2003); and, a hallmark of recent research on models and modeling is its optimism about the possibility of helping average ability students exhibit extraordinary achievements involving deeper and higher-order mathematical thinking (Lesh & Doerr, 2002). However, when we recognize that there is a need to develop mathematics out of situations that are meaningful to students, we recognize that, for "pure" mathematicians, "pure" mathematical systems are meaningful—and that many can be made so for youngsters in school. On the other hand, for children who have not yet developed most of these systems to a level that they can be explored meaningfully, there is absolutely no danger that mathematics textbooks are likely to include too many activities that could be called "model-eliciting" (Lesh et al., 2000). In fact, in classes that we teach for experienced teachers, when we give participants the assignment to review their own textbooks to try to find instances of problem-solving activities that satisfy the six design principles that define what it means to be a model-eliciting activity (Lesh et al., 2000), they typically discover that no such problems are included. That is, every problem violates every one of the six principles! Of course, model-eliciting activities are not the only kinds of problems that may be beneficial to include in instruction. Nonetheless, since model-eliciting activities were designed explicitly to focus on mathematical ideas and abilities that are needed for success beyond school in a technology-based age of information, it is noteworthy that current textbooks and tests typically include no such activities.

Concerning Relations Between Basic Skills and More Powerful Constructs

Models and modeling perspectives should not be interpreted as advocating the neglect of basic facts and skills. This is true for the same reasons that coaches need not neglect fundamentals and skills simply because their teams are allowed to scrimmage occasionally. In other publications, we have dealt extensively with this issue of attaining a balance between skill development and the development of deeper and higher-order abilities (Dark, 2003; Lesh, 2001). For now, we simply observe that, if a teacher's goal is to help students develop more powerful ways of thinking about that are known to be difficult to understand—such as ideas related to density, rates of growth, forces, sampling, and chance—then more is needed than simply introducing a few new facts and skills. All of these ideas presuppose the development of some specialized conceptual system; and, unless students are challenged to express their underlying ways of thinking about these constructs in forms that create the need for testing and revision, new facts and skills tend to be grafted onto inappropriate models for making sense of experiences. Our research suggests that activities that challenge students to express their ways of thinking in the form of complex artifacts provide some of the most effective instructional "leverage points" for developing powerful conceptual tools that radically amplify their mathematical abilities (Kaput & Schorr, in press; Lehrer & Schauble, 2002; Lesh & Kelly, 2002).

ACKNOWLEDGMENTS

The research described in this article was sponsored partly by the University of Wisconsin's National Center for the Improvement of Science Learning and Achievement in Mathematics and the Sciences, as well as by the Lucent Educational Foundation and the AT&T Education Foundation through Purdue's Center for Twenty-first Century Conceptual Tools. Conclusions represent the views of the authors and do not reflect those of supporting institutions.

REFERENCES

Beth, E., & Piaget, J. (1966). *Mathematical epistemology and psychology*. Dordrecht, The Netherlands: Reidel.

Bruner, J. (1963). *The process of education*. New York: Vantage Books.

Carlson, M., Larsen, S., & Lesh, R. (2003). Integrating a models and modeling perspective with existing research and practice. In R. Lesh & H. M. Doerr (Eds.) *Beyond constructivism: Models and modeling perspectives on mathematics teaching, learning, and problem solving* (pp. 465–478). Mahwah, NJ: Lawrence Erlbaum Associates, Inc.

Carpenter, T. P., Fennema, E., & Romberg, T. A. (1993). *Rational numbers: An integration of research*. Hillsdale, NJ: Lawrence Erlbaum Associates, Inc.

Clark Koellner, K., & Lesh, R. (2003). A modeling approach to describe teacher knowledge. In R. Lesh, & H. M. Doerr (Eds.), *Beyond constructivism: Models and modeling perspectives on mathematics teaching, learning, and problem solving* (pp. 159–173). Mahwah, NJ: Lawrence Erlbaum Associates, Inc.

Cobb, P., & McClain, K. (2001). An approach for supporting teachers' learning in social context. In F.-L. Lin & T. Cooney (Eds.), *Making sense of mathematics teacher education* (pp. 207–232). Dordrecht, The Netherlands: Kluwer.

Cobb, P., & Yackel, E. (Eds.). (1998). *Symbolizing, communicating, and mathematizing*. Mahwah, NJ: Lawrence Erlbaum Associates, Inc.

Cramer, K. (2003). Using a translation model for curriculum development and classroom instruction. In R. Lesh & H. M. Doerr (Eds.), *Beyond constructivism: Models and modeling perspectives on mathematics teaching, learning, and problem solving* (pp. 449–463). Mahwah, NJ: Lawrence Erlbaum Associates, Inc.

Dark, M. (2003). A models and modeling perspective on skills for the high performance workplace. In R. Lesh & H. M. Doerr (Eds.), *Beyond constructivism: Models and modeling perspectives on mathematics teaching, learning, and problem solving* (pp. 279–293). Mahwah, NJ: Lawrence Erlbaum Associates, Inc.

Dewey, J. (1982). *Pragmatism; The classic writings* (H.S. Thayer, Ed., pp. 253–334). Indianapolis, IN: Hackett.

Dewey, J. (1990). *John Dewey's pragmatic technology* (L. Hickman, Ed.). Bloomington: Indiana University Press.

Dewey, J. (1997). *Pragmatism: A reader* (L. Menand, Ed., pp. 182–271). New York: Vintage Books.

Dewey, J. (1998). *Transforming experience* (M. Eldridge, Ed.). Nashville, TN: Vanderbilt University Press.

diSessa, A. A. (1988). Knowledge in pieces. In G. Forman & P. B. Pufall (Eds.), *Constructivism in the computer age* (pp. 49–70). Hillsdale, NJ: Lawrence Erlbaum Associates, Inc.

Doerr, H. M., & Lesh, R. (2003). A modeling perspective on teacher development. In R. Lesh & H. M. Doerr (Eds.), *Beyond constructivism: Models and modeling perspectives on mathematics teaching, learning, and problem solving* (pp. 125–140). Mahwah, NJ: Lawrence Erlbaum Associates, Inc.

Freudenthal, H. (1973). *Mathematics as an educational task.* Dordrecht, Netherlands: Reidel.

Gee, J. P. (1997). Thinking, learning, and reading: The situated sociocultural mind. In D. Kirshner & J. A. Whitson (Eds.), *Situated cognition: Social, semiotic, and psychological perspectives* (pp. 235–259). Mahwah, NJ: Lawrence Erlbaum Associates, Inc.

Greeno, J. (1991). Number sense as situated knowing in a conceptual domain. *Journal for Research in Mathematics Education,* 22(3), 170–218.

Harel, G., & Lesh, R. (2003). Local conceptual development of proof schemes in a cooperative learning setting. In R. Lesh & H. M. Doerr (Eds.), *Beyond constructivism: Models and modeling perspectives on mathematics teaching, learning, and problem solving* (pp. 405–431). Mahwah, NJ: Lawrence Erlbaum Associates, Inc.

James, W. (1982). *Pragmatism; The classic writings* (H. S. Thayer, Ed., pp. 123–250). Indianapolis, IN: Hackett.

Johnson, T., & Lesh, R. (2003). A models and modeling perspective on technology-based representational media. In R. Lesh & H. M. Doerr (Eds.), *Beyond constructivism: Models and modeling perspectives on mathematics teaching, learning, and problem solving* (pp. 265–277). Mahwah, NJ: Lawrence Erlbaum Associates, Inc.

Kaput J. (1994a). Democratizing access to calculus: New routes to old roots. In A. H. Schoenfeld (Ed.), *Mathematical thinking and problem solving.* Hillsdale, NJ: Lawrence Erlbaum Associates, Inc.

Kaput, J. (1994b). The representational roles of technology in connecting mathematics with authentic experience. In R. Bieler, R. W. Scholz, R. Strasser, & B. Winkelman (Eds.), *Mathematics didactics as a scientific discipline.* Dordrecht, The Netherlands: Kluwer.

Kaput, J. (in press). Overcoming physicality and the eternal present: Cybernetic manipulatives. In R. Sutherland & J. Mason (Eds.), *Visualization and technology in mathematics education.* New York: Springer-Verlag.

Kaput, J. J., & Schorr, R. Y. (in press). Changing representational infrastructures changes most everything: The case of SimCalc, algebra and calculus. In K. Heid & G. Blume (Eds.), *Research on the impact of technology on the teaching and learning of mathematics.* Retrieved from http://www.simcalc.umassd.edu/NewWebsite/downloads/ChangingInfrastruct.pdf

Kelly, A., & Lesh, R. (Eds.). (2000a). *The handbook of research design in mathematics and science education.* Hillsdale, NJ: Lawrence Erlbaum Associates, Inc.

Kelly, A., & Lesh, R. (2000b). Trends and shifts in research methods. In A. Kelly & R. Lesh (Eds.), *The handbook of research design in mathematics and science education.* Hillsdale, NJ: Lawrence Erlbaum Associates, Inc.

Lehrer, R., & Schauble, L. (2000a). Inventing data structures for representational purposes: Elementary grade students' classification models. *Mathematical Thinking and Learning, 2,* 51–74.

Lehrer, R., & Schauble, L. (2000b). Model-based reasoning in mathematics and science. In R. Glaser (Ed.), *Advances in instructional psychology (Vol. 5).* Mahwah, NJ: Lawrence Erlbaum Associates, Inc.

Lehrer, R., & Schauble, L. (2003). Origins and evolutions of model-based reasoning in mathematics and science. In R. Lesh & H. M. Doerr (Eds.), *Beyond constructivism: Models and modeling perspectives on mathematics teaching, learning, and problem solving* (pp. 59–70). Mahwah, NJ: Lawrence Erlbaum Associates, Inc.

Lesh, R. (2001). Beyond constructivism: A new paradigm for identifying mathematical abilities that are most needed for success beyond school in a technology based age of information. In M. Mitchelmore (Ed.), *Technology in mathematics learning and teaching: Cognitive considerations: A special issue of the Mathematics Education Research Journal.* Melbourne, Australia: Australia Mathematics Education Research Group.

Lesh, R. (2002). Research design in mathematics education: Focusing on design experiments. In L. D. English (Ed.), *Handbook of international research in mathematics education* (pp. 27–50). Hillsdale, NJ: Lawrence Erlbaum Associates, Inc.

Lesh, R., & Carmona, G. (2003). Piagetian conceptual systems and models for mathematizing: Everyday experiences. In R. Lesh & H. M. Doerr (Eds.), *Beyond constructivism: Models and modeling perspectives on mathematics teaching, learning, and problem solving* (pp. 71–96). Mahwah, NJ: Lawrence Erlbaum Associates, Inc.

Lesh, R., & Doerr, H. (1998). Symbolizing, communicating, and mathematizing: Key components of models and modeling. In P. Cobb & E. Yackel (Eds.), *Symbolizing, communicating, and mathematizing.* Mahwah, NJ: Lawrence Erlbaum Associates, Inc.

Lesh, R., & Doerr, H. M. (Eds.). (2003a). *Beyond constructivism: Models and modeling perspectives on mathematics teaching, learning, and problem solving.* Mahwah, NJ: Lawrence Erlbaum Associates, Inc.

Lesh, R., & Doerr, H. M. (2003b). Foundations of a models and modeling perspective on mathematics teaching, learning, and problem solving. In R. Lesh, & H. M. Doerr (Eds.), *Beyond constructivism: Models and modeling perspectives on mathematics teaching, learning, and problem solving* (pp. 3–33). Mahwah, NJ: Lawrence Erlbaum Associates, Inc.

Lesh, R., & Doerr, H. M. (2003c). In what ways does a models and modeling perspective move beyond constructivism? In R. Lesh & H. M. Doerr (Eds.), *Beyond constructivism: Models and modeling perspectives on mathematics teaching, learning, and problem solving* (pp. 519–556). Mahwah, NJ: Lawrence Erlbaum Associates, Inc.

Lesh, R., & Heger, M. (2003). What mathematical abilities are needed for success beyond school in a technology-based age of information? In M. Thomas (Ed.), *Technology in mathematics education.* New Zealand: University of Auckland Press.

Lesh, R., Hoover, M., Hole, B., Kelly, A., & Post, T. (2000). Principles for developing thought-revealing activities for students and teachers. In A. Kelly & R. Lesh (Eds.). *Handbook of research design in mathematics and science education* (pp. 591–646). Mahwah, NJ: Lawrence Erlbaum Associates, Inc.

Lesh, R., Zawojewski, J. S., & Carmona, G. (2003). What mathematical abilities are needed for success beyond school in a technology-based age of information? In R. Lesh & H. M. Doerr (Eds.), *Beyond constructivism: Models and modeling perspectives on mathematics teaching, learning, and problem solving* (pp. 205–222). Mahwah, NJ: Lawrence Erlbaum Associates, Inc.

Lester, F. K., & Kehle, P. (2002). From problem solving to modeling: The evolution of thinking about research on complex mathematical activity. In R. Lesh & H. M. Doerr (Eds.), *Beyond constructivism: Models and modeling perspectives on mathematics teaching, learning, and problem solving* (pp. 501–517). Mahwah, NJ: Lawrence Erlbaum Associates, Inc.

Lesh, R., Lester, F. K., & Hjalmarson, M. (2003). A models and modeling perspective on metacognitive functioning in everyday situations where problem solvers develop mathematical constructs. In R. Lesh & H. M. Doerr (Eds.), *Beyond constructivism: Models and modeling perspectives on mathematics teaching, learning, and problem solving* (pp. 383–402). Mahwah, NJ: Lawrence Erlbaum Associates, Inc.

McClain, K. (2003). Task-analysis cycles as tools for supporting students' mathematical development. In R. Lesh & H. M. Doerr (Eds.), *Beyond constructivism: Models and modeling perspectives on mathematics teaching, learning, and problem solving* (pp. 175–189). Mahwah, NJ: Lawrence Erlbaum Associates, Inc.

Mead, G. H. (1910). Social consciousness and the consciousness of meaning. *Psychological Bulletin, 7,* 397–405.

Middleton, J., Lesh, R., & Heger, M. (2003). Interest, identity, and social functioning: Central features of modeling activity. In R. Lesh, & H. M. Doerr (Eds.), *Beyond constructivism: Models and modeling perspectives on mathematics teaching, learning, and problem solving* (pp. 405–431). Mahwah, NJ: Lawrence Erlbaum Associates, Inc.

Minsky, M. (1987). *The society of mind.* New York: Simon & Schuster.

Olson, L. (1994). Algebra focus of trend to raise stakes. *Education Week, 1*(11).

Pickering, A. (1995). *The mangle of practice.* Chicago: University of Chicago Press.

Peirce, C. S. (1982). In H. S. Thayer (Ed.), *Pragmatism: The classic writings* (pp. 43–120). Indianapolis, IN: Hackett.

Schoenfeld, A. H. (1982). Some thoughts on problem-solving research and mathematics education. In F. K. Lester & J. Garofalo (Eds.), *Mathematical problem solving: Issues in research* (pp. 27–37). Philadelphia: Franklin Institute Press.

Schorr, R. Y., & Lesh, R. (2003). A modeling approach for providing teacher development. In R. Lesh & H. M. Doerr (Eds.), *Beyond constructivism: Models and modeling perspectives on mathematics teaching, learning, and problem solving* (pp. 141–157). Mahwah, NJ: Lawrence Erlbaum Associates, Inc.

Shternberg, B., & Yerushalmy, M. (2003). Models of functions and models of situations: On the design of modeling-based learning environments. In R. Lesh & H. M. Doerr (Eds.), *Beyond constructivism: Models and modeling perspectives on mathematics teaching, learning, and problem solving* (pp. 479–498). Mahwah, NJ: Lawrence Erlbaum Associates, Inc.

vanReeuwijk, M., & Wijers, M. (2003). Explanations why? The role of explanations in answers to (assessment) problems. In R. Lesh & H. M. Doerr (Eds.) *Beyond constructivism: Models and modeling perspectives on mathematics teaching, learning, and problem solving* (pp. 191–202). Mahwah, NJ: Lawrence Erlbaum Associates, Inc.

Vygotsky, L. S. (1978). *Mind in society: The development of higher psychological processes.* Cambridge, MA: Harvard University Press.

Wertsch, J. (1985). *Vygotsky and the social formation of mind.* Cambridge, MA: Harvard University Press.

Wertsch, J. (1991). *Voices of the mind: A sociocultural approach of mediated action.* London: Harvester.

Zawojewski, J. S., & Bowman, K. (2001–2004). Small group mathematical modeling approaches to improved gender equity in engineering project (SGMM). National Science Foundation funded grant.

Zawojewkski, J. S., & Lesh, R. (2003). A models and modeling perspective on problem solving. In R. Lesh & H. M. Doerr (Eds.), *Beyond constructivism: Models and modeling perspectives on mathematics teaching, learning, and problem solving* (pp. 317–336). Mahwah, NJ: Lawrence Erlbaum Associates, Inc.

Zawojewski, J. S., Lesh, R., & English, L. (2003). A models and modeling perspective on the role of small group learning activities. In R. Lesh & H. M. Doerr (Eds.), *Beyond constructivism: Models and modeling perspectives on mathematics teaching, learning, and problem solving.* Mahwah, NJ: Lawrence Erlbaum Associates, Inc.

MATHEMATICAL THINKING AND LEARNING, *5*(2&3), 131–156

Structuring Error and Experimental Variation as Distribution in the Fourth Grade

Anthony J. Petrosino
Curriculum and Instruction
The University of Texas at Austin

Richard Lehrer and Leona Schauble
Department of Teaching and Learning
Peabody College, Vanderbilt University

Humans appear to have an inborn propensity to classify and generalize, activities that are fundamental to our understanding of the world (Medin, 1989). Yet, however one describes objects and events, their variability is at least as important as their similarity. In *Full House*, Stephen Jay Gould neatly drove home this point with his choice of a chapter subhead: "Variability as Universal Reality" (1996, p. 38). Gould (1996) further noted that modeling natural systems often entails accounting for their variability.

An example now widely familiar from both the popular and professional press (e.g., Weiner, 1994) is the account of how seemingly tiny variations in beak morphology led to dramatic changes in the proportions of different kinds of Galapagos finches as the environment fluctuated over relatively brief periods of time. Understanding variability within and between organisms and species is at the core of grasping a big idea like "diversity" (NRC, 1996). Yet, given its centrality, variability is given very short shrift in school instruction. Students are given few, if any conceptual tools to reason about variability, and even if they are, the tools are rudimentary, at best. Typically, these tools consist only of brief exposure to a few statistics (e.g., for calculating the mean or standard deviation), with little focus on the more encompassing sweep of *data modeling*. That is, stusdents do not typically participate in contexts that allow them to develop questions, consider qualities of measures and attritubutes relevant to a question, and then go on to structure data and make inference about their questions.

Requests for reprints should be sent to Anthony Petrosino, Curriculum and Instruction, The University of Texas at Austin, 1912 Speedway, SZB 462–A, Austin, TX 78712. E-mail: ajpetrosino@mail.utexas.edu

Our initial consideration of this problem suggested that distribution could afford an organizing conceptual structure for thinking about variability located within a more general context of data modeling. Moreover, we conjectured that characteristics of distribution, like center and spread, could be made accessible and meaningful to elementary school students if we began with measurement contexts where these characteristics can be readily interpreted as indicators of a "true" measure (center) and the process of measure (spread), respectively (Lehrer, Schauble, Strom, & Pligge, 2001). Our conjecture was that with good instruction, students could put these ideas to use as resources for experiment, a form of explanation that characterizes many sciences.

Accordingly, here we report an 8-week teaching and learning study of elementary school students' thinking about distribution, including their evolving concepts of characteristics of distribution, like center, spread, and symmetry. The research was conducted with an intact class of fourth-grade students and their teacher, Mark Rohlfing. The context for these investigations was a series of tasks and tools aimed at helping students consider error as distributed and as potentially arising from multiple sources. For students to come to view variability as distribution, and not simply as a collection of differences among measurements, we introduced distribution as a means of displaying and structuring variation among their observations of the "same" event.

This introduction drew from the historical ontogeny of distribution, which suggests that reasoning about error variation may provide a reasonable grounding for instruction (Konold & Pollatsek, 2002; Porter, 1986; Stigler, 1986), and from our previous research with fifth-grade students who were modeling density of various substances (Lehrer & Schauble, 2000; Lehrer et al., 2001). In our previous research, students obtained repeated measurements of weights and volumes of a collection of objects, and discovered that measurements varied from person-to-person and also, by method of estimation. For example, students found the volumes of spheres with water displacement, the volume of cylinders as a product of height and an estimate of the area of the base, and the volume of rectangular prisms as a product of lengths. Differences in the relative precision of these estimates were readily apparent to students (e.g., measures obtained by water displacement were much more variable than those obtained by product of lengths). Also evident were differences in measurements across measurers (e.g., one student's estimate of 1/3 of a unit of area might be taken as ½ of a unit by a different student). Students estimated true values of each object's weight and volume in light of variation by recourse to an invented procedure analogous to a trimmed mean (extreme values were eliminated and the mean of the remaining measures was found as the "best guess" about its value). In summary, this previous research suggested that contexts of measurement afforded ready interpretation of the center of a distribution as an indicator of the value of an attribute and variation as an indicator of measurement process. However, we did not engage students in examining the structure of variation (e.g., its potential symmetry), nor did we investigate sources of variation other than person and instrument.

Consequently, in this 8-week teaching study, we engaged students in making a broader scope of attribution about potential sources and mechanisms that might produce variability in light of the processes involved in measuring. Several items were measured; they were selected for their potential for highlighting these ideas about variability and structure. Students measured first, the lengths of the school's flagpole and a pencil, and later, the height of model rockets at the apex of their flight. Multiple sources of random error were identified and investigated. To distinguish between random and systematic variation, students conducted experiments on the design of the rockets (rounded vs. pointed nose cones). The goal of these investigations was to determine whether the differences between the resulting distributions in heights of the rockets were consistent with random variation, or instead, if the difference could be attributed to the shape of the nose cone. During their investigations students identified and discussed several potential contributions to measurement variability (including different procedures for measuring, precision of different measurement tools, and trial by trial variability). The expectation that there was a "true" measure served to organize students' interpretations of the variability that they observed (in contrast to the interpretation of inherent variability, for example, in the heights of people, a shift that has historically been difficult to make—see Porter, 1986).

Without tools for reasoning about distribution and variability, students find it impossible to pass beyond mere caricatures of inquiry, which at present, dominate much of the current practice in science instruction. Consider, for example, a hypothetical investigation about model rocket design. Students launch several rockets that have rounded nose cones and measure the height of each rocket at the apex of its flight. Next they measure the launch heights of a second group of rockets with pointed nose cones. The children find, not surprisingly, that the heights are not identical. Yet, what do these results mean? How different must the results be to permit a confident conclusion that the nose cone shape matters?

Much current science instruction fails to push beyond simple comparison of outcomes. If the numbers are different, the treatments are presumed to differ, a conclusion, or course, that no practicing scientist would endorse. Because most students do not have the tools for understanding ideas about sampling, distribution, or variability, teachers have little recourse, but to stick closely to "investigations" of very robust phenomena that are already well understood. The usual result is that students spend their time "investigating" a question posed by the teacher (or the textbook), following procedures that are defined in advance. This kind of enterprise shares some surface features with the "inquiry" valued in the science standards, but not its underlying motivation or epistemology. For example, students rarely participate in the development, evaluation, and revision of questions about phenomena that interest them personally. Nor do they puzzle through the challenges of instrumentation, what Pickering (1995) described as achieving a "mechanic grip" on the world—for example, by deciding how to parse the world into measurable attributes and then coming to agreement about best ways of measuring them. Many teachers pursue "canned" investigations not for lack of enthusiasm for this kind of genuine inquiry, but because encouraging students to develop and investigate their own questions requires that they have the requisite tools for interpreting the results.

In the study we report here, teachers and students worked with researchers to engage in coordinated cycles of (a) planning and implementing instruction to put these tools in place, and (b) studying the forms of student thinking—both resources and barriers—that emerged in the classroom. Although presented here as separate phases, these two forms of activity were, in fact, concurrent and coordinated. The development of student thinking was conducted in the context of particular forms of instruction; and in turn, instructional planning was continually guided by and locally contingent on what we (teacher and researchers) were learning about student thinking.

METHOD

Participants

Participants were 22 fourth grade students (12 boys, 10 girls) from an intact fourth grade class in a public elementary school located in the Midwest. The students and their teacher were part of a school-wide reform initiative aimed at changing teaching and learning of mathematics and science (for details on this initiative, see Lehrer & Schauble, 2000). The teacher had 11 years of teaching experience at the time of this study. However, neither the students nor the teacher had any previous experience with the forms of mathematics and science pursued in this study.

Procedure

The 22 students participating in the research met as a class with their teacher and the first author, with occasional interactions with the second author, over an 8-week period (19 sessions, April through May 1999) in blocks of time ranging from 50–90 min. During each session, students worked in groups of 4–5 at large round tables in the classroom. The teacher often moved from whole group to small group to individual instruction, along with variations on these configurations. This cycle of varying participatory structure, which was part of the preexisting classroom culture, was retained throughout the course of the study. However, most of what we report occurred in the context of whole-group discussions. The major sources of data were field notes, transcribed audiotapes of student interviews, videotapes, as well as the inscriptions of data that the students and teacher created and manipulated throughout the study.

At the conclusion of instruction, the whole class responded in writing to seven assessment items. The first two were slight modifications of NAEP (National Assessment of Educational Progress) released items designed to assess graphic interpretation skill. The remaining five items were adapted from or borrowed from Paul Cobb and Kay McClain's work on statistical reasoning with middle-grades students (Cobb, 1999; Cobb, McClain, & Gravemeiser, in press).

We also administered a clinical interview designed to probe student reasoning more extensively. The first item in this interview probed student conceptions of experiment, especially the grounds for knowing whether one could be confident about the effect of a variable (e.g., the shape of the nose cone of a rocket). The second item was developed by Paul Cobb and Kay McClain to assess middle school students' understanding of center of distribution and the role it plays when comparing distributions (McGatha, 2000). The third item probed student understanding of measurement variation in light of a true score. Here we noted whether students thought that measurement variation would be relatively symmetric and centered around the true score. Student thinking about the effects of instrumentation on distribution was assessed by comparing their conjectures about these distributions in light of what they knew about the precision of each of two instruments used during instruction (height-o-meter and AltiTrak™). The final item assessed student strategies for comparing distributions resulting from different experimental conditions.

Materials

Most materials were everyday "off-the-shelf" objects like rulers and string. Yet, because we wished students to consider variation among observations due to instrumentation, students also either constructed or were provided more specialized measurement devices. These included a cardboard "height-o-meter" (see Figure 1) that worked in a manner similar to a commercially produced tool, the AltiTrak™ (Estes Industries 1995 Catalog number 2232). Both the height-o-meter and the AltiTrak™ finder rely on properties of right triangles for calculating heights. However, these devices differ in the precision of their estimations of height, because of the comparative rigidity of materials (cardboard vs. plastic) and the tolerances of hand-crafted vs. machined components.

Students also constructed model rockets from kits. The rocket kit was a product of Estes Industries, and the model was known as The Big Bertha™. The rockets measured 24 inches in length with a diameter of 1.637 inches and a weight of 2.2 ounces. The Big Bertha™ model was chosen for a number of reasons, including its ease of handling, prior success with similar-aged students, and fundamentally sound and durable construction characteristics (Petrosino, 1995). The model has changeable nose cones and fins, making modification a fairly simple process. The flexibility of modification to the initial design of the rockets was a crucial consideration in the selection of The Big Bertha™ model. In these modifications, two different types of nose cones were used. One was the standard rounded nose cone that comes with The Big Bertha™ kit (Estes Industries 1995 Catalog number 3166, NC–60b), and the other was a pointed cone (Estes Industries 1995 Catalog number 3165, NC–60a).

FIGURE 1 Commerical and Hand-Crafted Tools for Measuring Height. Left: Paper-measuring device used by students to initially measure school ground flagpole. Right: Altitrak, professional device used by students to remeasure school ground flagpole and subsequent model rocket launches.

The rocket motor (or engine) is the device inserted into a model rocket to impart a thrust force to the rocket, boosting it into the air. Model rocket motors are miniature solid fuel rocket engines that contain propellant, a delay element, and an ejection charge. The students used B4–2 single-stage model rocket motors (Estes Industries 1995 Catalog number 1601) for the propulsion of the model rockets. These motors have a maximum trust of 13.3 Newtons and a thrust duration of 1.2 sec. The B4–2 typically has a burnout altitude of 52m (170 feet) and a peak altitude of 295m (950 feet) under ideal conditions using specially designed model rockets. However, heights are significantly reduced by the use of large model rockets such as The Big Bertha™. Through previous experience (Petrosino, 1998), we knew that these motors would provide adequate height to differentiate the effects of nose cones from random trial-to-trial variation.

SEQUENCE OF INSTRUCTION

Here we describe the series of tasks and tools that we designed to lead students to consider error as distributed and as potentially arising from multiple sources. In the context of these tasks, distribution was introduced as a means of displaying and structuring variation among observations. Students were engaged in making attributions about potential sources and mechanisms that might produce variability in light of this structure. Sources of random error included individual differences among measurers (all activities), instrument variation (paper vs. plastic measuring instruments), and replication variation (multiple trials of rocket launches). To distinguish between random and systematic variation, students conducted experiments on the design of rockets (rounded vs. pointed nose cones). The purpose was to determine whether differences between the resulting distributions of rocket heights (obtained with round vs. pointed nose cones) were consistent with random variation, or if the differences perhaps could be attributed to the design of the rockets (the independent variable under investigation). In the following sections, we identify some key milestones in the children's understanding of distribution and error as they emerged during the instructional sequence, and ultimately, how students appropriated these conceptual tools for purposes of experiment.

Structuring Variation in Measure as a Distribution

All students measured the height of their school's flagpole using a hand-made "height-o-meter." We expected this procedure to generate a set of measures consistent with a Gaussian (i.e., normal) distribution. The purpose was to make characteristics of center and spread of distribution salient. First, we expected that students would readily agree that the flagpole had a "true" height and therefore, the center of the distribution might be interpreted as its (approximate) measure. Second, the make-shift nature of their measuring instruments might inspire ideas about why one would expect some measurements above and below the actual height of the flagpole. Symmetry in measure could then serve later as a potential explanation for symmetries in the shape of the distribution of their measurements. Third, we hoped that it might occur to some students to group similar values of measurements. These groups could serve as an introduction to the important mathematical idea of a distribution as a function of the density of the number of observations within particular intervals. These intervals are "vanishingly small" in the language of calculus, but because this mathematical tool was not available to our students, we intended to provoke consideration of intervals as groups of cases that could be regarded as similar.

The measurements that students recorded ranged from 6.2 m to 15.5 m, with a median value of 9.7 m. Students wrote their measures on 3 × 5 inch index cards, one card for each measurement taken by one student-measurer.

Inventing displays of measurements. Working in small groups, students were challenged to arrange the cards to show their best sense of "how you would organize this," in the words of their teacher, Mr. Rohlfing. Our intention was to set the stage for more conventional displays by helping students understand their purposes. As expected, students' displays of these data varied, and the variation provided an opportunity to develop a sense of how different representations made some aspects of the data more or less visible.

Many of the displays invented by students were apparently generated to conform to the size and shape of the tables where they worked. For example, one group organized the data in a "wheel." In this circular configuration, values were ordered so the number directly opposite the lowest value was designated as the "middle number" (the median). Students in this group reported that they favored this display because it enabled them to tell at a glance how many numbers there were above and below the "middle number." As Mr. Rohlfing traveled from group to group, he highlighted this notion of the middle number as a way of dividing the batch of data into two equal groups. Other groups created intervals of ten and referred to them as "bins." Mr. Rohlfing pointed out to this group that the resulting display helped them see at a glance that "there were a lot more 9's" (e.g., measurements in the 9" bin than in any of the surrounding bins). This teacher move helped establish that despite differences, values of data could be considered as similar.

Mr. Rohlfing next orchestrated a collective discussion of the displays created by the small groups. During this discussion, he oriented conversation toward considering the idea of interval (the bins) and evaluating extreme measures, like 15.5. He first ordered the data from smallest to largest, horizontally on the board. He then used blue index cards to mark column bins (10's) and pink cards to represent data values. The resulting display is represented in Figure 2. Mr. Rohlfing used this display to provoke the possibility that perhaps not all the measurements were equally likely. To raise this issue, he focused student attention on the value of 15.5, on the far right tail of the display.

Katie was the first to express concern about the variability in the class's measurements. She suggested, "Someone probably goofed with the numbers, 'cause there's such a big difference." Mr. Rohlfing noted that values that were very different from the others were called outliers. He followed up this comment by asking, "When does it become an outlier? What if I change it to 16? 17? 18? Raise your hand when you think it's an outlier."

Issac responded, "There's not much difference between 9.1 and 9.2, but there is a lot of difference between 13.8 and 15.5. It looks like there's more difference when you put them in columns." Issac's comment set the stage for further conversation about what bins allowed one to "see" and how without them, a value like 15.5 presented in an ordered list would not as apparently be an anomaly.

Mr. Rohlfing then asked students about their confidence in these measures, and especially asked them to compare measurements closer to "the middle" to those

6	7	8	9	10	11	12	13	14	15
62	7.8	87	91	10	11	123	13		155
64		88	92	102					
68			93	103					
			97	106					
			98						

FIGURE 2 Facsimile of Display of Measures of the Height of the Flagpole.

that they considered outliers. This teaching move focused student attention on the relation between the (emerging) qualities of the distribution and their acts of measurement. Keith noted that "….there was only one 15.5." Charlie followed up by suggesting, "9 comes up more than any other number, so I think that might be more accurate."

Students worked in small groups to concur on the measures that seemed to warrant most confidence, and the class reconvened shortly thereafter to reconsider this question. Student notions of confidence took a variety of forms, but many argued that more central values were the ones that were most trustworthy. For example, one group of students proposed, "We are 95% sure it is not 15.5. It's more likely that the number would be between 9.1 and 10, 'cause more people [cases] were there. A good majority were there, and the odds were really good that the number would be there." Another group proposed ratings of confidence and suggested a rating of 9 for those in the 9 bin and a rating of 2 for 6.2. Ratings like these were generated to indicate qualitative confidence in particular measures (in relation to a true height) and were not coupled to proportions or other quantities.

Students generally thought that more measurements within a bin were an indication that the actual flagpole height was somewhere in that bin, and that the true height was less likely to be in bins with fewer cases. Some groups, however, wished to temper these conclusions. For example, one suggested, "There are more 9's than anything, but there's also a good chance that any other number might be it, too. So, we did it on percent, 100% being absolutely certain and 0% meaning nothing. 15.5 = 2%. For 10 we have 35%." This comment opened the possibility of a middle region encompassing values in the 9- and 10-meter intervals. Some students suggested that this middle region resulted from the measurement process itself, which would create values "a little over" or a "little under." Others suggested that extreme values represented being "a lot over or under," and thus were less likely to occur.

Accentuating Precision of Measure

To accentuate relations between measurement error and the resulting characteristics of distribution, students next used a ruler to measure a regular No. 2 pencil. We intended to highlight the effects of precision of measure on the "spread" of the resulting observations. As mentioned previously, we expected that students might begin to think about the inherent variation due to instrumentation. A second goal was to make problematic the fact that in spite of the greater precision of the ruler measures (compared to height-o-meters), it remained the case that the measures were still not all identical.

Students worked collectively to put their measures of the length of the pencil into a group display. Students decided again to create bins (intervals), and found that nearly all cases were tightly clumped around 18 cm. However, there were a series of outliers (7.07, 7.3,7.6, 7.7, 7.7) that attracted attention by virtue of their extreme distance from the rest of the measures. One student accounted for the outliers by demonstrating that values in the "7 clump" arose from using different units of measure (i.e., the other side of the ruler). Several students again pointed out that they felt more confident when others produced the same measure and that the measures were less "spread out" than previously. Earle suggested another source of "human" error, noting that variations in student height might have affected the use of height-o-meter to measure the flagpole, but for the pencil, "It didn't matter how tall you were."

Typical measure. Mr. Rohlfing asked students to identify a "typical" measure of the length of the pencil. Several students proposed that a median value represented the "middle" and thus was a best guess about true height, because estimates tended to be a little over or under. Other students suggested that the mode might be a better indicator because it indicated agreement about a value reached by different measurers, which would be hard to accomplish unless they were really measuring the same length. Because the median and mode were the same in this batch of data, both justifications were accepted. Several students proposed constructing an average, but their suggestion was not considered further, perhaps because the median and mode were perceived as clear indicators of the middle of the distribution by most of the class.

Spread as a distribution of difference. Mr. Rohlfing then asked students to compare the relative precision of the flagpole and pencil measurements. Students gestured with their hands to indicate the relative compactness of the intervals in each display, suggesting that the difference between the measures was obvious. Their teacher raised the ante by pressing students to consider how much more precise each measurer was in the pencil context, compared to the flagpole context. At this point, one of the researchers (AP) asked "what a spread of zero might look like," tacitly suggesting that spread could be quantified. Students replied that zero would imply uniformity of measure: "All got the same number." Will added that this would mean that everyone obtained the "typical" number (the center).

Adopting the perspective of a measurer, the teacher initiated discussion about the difference between each measurement and the center for the entire set of pencil data. He began by introducing the use of a signed quantity to differentiate between under- and over-estimates. One of us (RL) introduced the metaphor of a difference as a "distance and direction you have to travel" to get from the center of the distribution to the value in question. The class readily established that in this case, negative distances corresponded to underestimates and positive distances to overestimates. Students then worked in small groups to calculate distances from the median for the flagpole and pencil measurements. They worked in a large group to make bins of difference. The resulting distributions of difference scores were examined, with intervals of one (e.g. –1 to –2, etc.) These distributions recapitulated the sense of relative compactness observed in the original measures, with many students again noting that the flagpole differences were "really spread out" compared to the pencil differences. Some students attempted to quantify their perception of relative spread numerically with a subjective rating scale (e.g., "5 would be a lot"), and Mr. Rohlfing seized this opportunity to generate discussion about how to quantify spread.

Spread numbers. Mr. Rohlfing suggested ordering the differences by magnitude only, rather than by magnitude and direction. "I'm gonna ignore what direction it is, and I'm just gonna look how far away it is"[from the median value]. Some students believed that the sign would affect the distance of the "travel numbers," so several cases were compared explicitly. For example, Mr. Rohlfing asked: "So, which is closer, the guy who got 9.2 (from a middle value of 9.7) or the guy who got 10.2?" Students readily observed: "Neither!" Finally, he introduced the notion of a "typical" spread or "spread number" by finding the median of this distribution of differences. This was a new procedure for students, so they worked in small groups to consider differences and then spread numbers for first the pencil and then the flag pole data.

While exploring this procedure for quantifying spread, some students suggested that they preferred averages as indicators of spread, rather than medians. Mr. Rohlfing exploited this suggestion by initiating exploration of the relative effects of outliers on means and medians. He suggested that students find the "middle" of 2, 3, 5 and 390. The result was a lengthy investigation of the qualities of means and medians, especially in light of the flagpole data with its more extreme outliers. Spread numbers were calculated as medians and as averages, and in both instances, were related to the relative spread of the two distributions of data. The context of measurement error afforded a ready interpretation of spread.

Mr. Rohlfing asked, "As your spread number gets higher, what does that tell you about the spread?"

Mitch replied, gesturing with his hands toward the distribution on the board, "The numbers are more spread out."

Mr. Rohlfing continued, "And as the spread numbers get lower…?"

Matt added, "Your spread will be small."

Mr. Rohlfing, however, had another question: "What does 1.6 mean?"

Kyle responded, "How far away the numbers were, like on average, from 9.7" (the median value).

Exploring alternatives to spread number. At this junction, several students suggested that a less cumbersome, more efficient means of indexing spread would be to find the range. Their teacher agreed, but suggested that the spread number was also a good way to think about a typical measurer. Students making the new proposal did not seem overly convinced by this perspective, so RL and Mr. Rohlfing created a contrasting case of "fake pole" data. This data set had the same range, median, and number of cases as the flagpole data, but was nearly bimodal.

Hence, even though it had the same range, its variability far exceeded that of the flag pole data. Mr. Rohlfing asked students how this fake pole data display looked different from the display of the flagpole data. "So something's happened. Looks like we have a lot over here and a lot over here {pointing to ends of distribution}. Our 15.5 {the outlier case in the right-hand tail of the distribution} is still the same. Where do they look clumpier?"

Mitch: "They're kind of clumping around 13 and 14."
Tanner: "6 and 7?"
Mr. Rohlfing: "Right, I see two big clumps here. Who can predict about the spread, how far away you are from this typical number?"
Mitch: "I think it will be bigger."

Julia agreed. "I think it will change, too, because when you clumped them, you changed them around. You added numbers in there that weren't there. There's more numbers that take up more space" (i.e., more of the numbers are more distant from the center). "We made it clumpy where it wasn't clumpy and made it not clumpy where it was clumpy. It's generally more separated. I think the {spread} number might be higher."

Working as a whole group, students created a distribution of difference scores for the fake-pole data, and Mr. Rohlfing juxtaposed this distribution with the distribution of differences obtained from the flagpole. Students readily contrasted the relative "clumpiness" of the two distributions. Once again, students worked in small groups to find the range and spread number of the fake pole data. The resulting exploration, which yielded a spread number of 3.3 for the fake pole data (as compared with 1.6 for the flagpole data) appeared to advance the notion that the spread number might be a way of characterizing spread in a wider variety of circumstances than the range alone.

Instrumentation and Spread

To take advantage of spread number as a conceptual tool for thinking about sources of measurement error, we asked students to measure the height of the flagpole with the AltiTrak™. Measurements with the AltiTrak™ proved far less variable than those taken earlier with the height-o-meter. In fact, there was a three-fold reduction in the value of the spread number. Several students attributed the reduction in variation to the "sturdiness" of the tool, noting that "it won't flop around." Mr. Rohlfing highlighted this source of measurement error and invited contrast to other sources. Students recalled "human" error (individual differences like "hand movement"), method variation (e.g., "failure to agree about where to start measuring" or "we're different heights"), and qualities of the object itself that might make it more easily measured (e.g., being able to grasp the pencil, but not the flagpole).

Putting Distributional Thinking in Service of Experiment

The concluding phase of instruction was designed to help students appreciate an experiment as a contrast between two distributions of values. Students first launched a model rocket with a rounded nose cone three times, with the caveat that the first launch would be discounted as "practice." They used their AltiTraks to record the apex of the rocket heights for each trial. As we later describe, these successive launches of the rounded nose cones set the stage for consideration of individual difference and trial variation as components of error. Next, students conducted "experimental" launches, launching a rocket with a pointed nose cone twice (launches 4, 5). They expected pointed nose cones to "cut through" the air, resulting in a higher altitude. Finally, they compared the resulting distributions to make a decision about whether rockets ought to be designed with rounded or pointed nose cones. (Other launches were made later to consider other comparisons that we do not focus on here.)

Constructing a reference distribution. Students displayed their measures of the altitude of the successive launches of the rocket with the rounded nose cone in "bins" of 10 m, ordered by height, with individual cases color-coded by order of launch (e.g., red for the first launch).

Noting several outliers among the measures of each launch, the class adopted the median rather than the mean as an indicator of "how high it really went." "Spread numbers" (medians and averages of the differences from the center) were also computed for each launch. Mr. Rohlfing asked: "Did the rocket go higher or lower the second time?" Students initially were drawn to the fact that a few values obtained during "launch 1" were greater than any value recorded for "launch 2." However, others were not convinced that the rocket really went higher at launch 1, noting that the color-coded cases were intermingled. Julia pointed out, "I think they're the same. Each launch has same median, 62." Carlie agreed: "There are three 62s on both. I'm pretty convinced that they did pretty equal. They are about the same."

Students did notice, however that the data distributions had different average "spread numbers" (the average spread numbers were 9.2 and 7.8, respectively). Mr. Rohlfing asked: "Does an average spread tell you how high it went?" Angeline said: "No." Kevin continued: "But you could determine which one was more accurate." When asked how he knew this, Kevin replied, "Because the average spread is lower, and there are no spots open" (i.e., no "holes" in the display of the data). Kevin's observation led to the more general conjecture that perhaps over the course of the investigation, the class was "getting better at measuring" with the AltiTrak™. To pursue this possibility, students sequentially ordered the average spread numbers for each launch: 12.3, 10.6, 9.2, 7.8, 7.3, 8.9. Mr. Rohlfing asked whether anyone could interpret the sequence: "So when someone asks, do you think the class got better at using AltiTrak, what would you tell her?"

Mitch replied, "I would tell her we did, because it did get better until this last one" (i.e., the last launch, where the average spread number was a bit higher).

Mr. Rohlfing asked, "Is that one [the final spread number] a lot farther away than the others?"

Mitch concluded, "It's not very far away from the rest. I'd still say we got better, because the majority..."

At this point, Abby interrupted, "I think we got better at it. From our first launch to our last launch, you can kind of see how we got better. This last one (last value) is not way up here."

Keith raised the possibility, however, that their improvement was decelerating from launch to launch: "I think we improved more the first time we launched. Here [between launch 1 and launch 2] we moved down 2. Here [between launch 2 and launch 3] we moved down 1." What Kevin had noticed, of course, was an example of a more general phenomenon, the practice effect.

One of us (AP) then asked, "I wonder if we would have gotten this same trend with the pencil?"

Matt thought about it for a minute, and replied, "I think we would get better, but not better like that. With the ruler you usually can't improve by a ton. It's such a small thing that you're measuring, it's hard to improve by a lot."

We took this discussion as evidence of a progressive objectification of data (Hancock, Kaput, & Goldsmith, 1992). That is, a summary statistic—the "spread number"—was now being used as an indicator of a process, rather than as a simple description.

Mr. Rohlfing suggested pooling the data from both launches with the rounded nose cones. Students found the median and typical spread number for these pooled data. Mr. Rohlfing next suggested a cut of the data into three "super-bins," one including all cases below the median and spread number, one encompassing the median with lower and upper bounds determined by the spread number, and one above the median and spread number. Students found that 51% of the 37 rounded nose cone measurements fell within the middle bin, 21% fell within the lower bin, and 27% within the upper bin. Mr. Rohlfing then posted the data for the pointed nose cone launches and asked students to predict "what if" the kind of nose cone mattered. Several students replied that they expected to see "clumps" of measures of pointed nose cone rocket heights in the upper bin of the reference distribution. However, to their surprise, when they portioned these data into the three "super-bins," 86% of the values fell into the lower bin and 14% in the middle bin. And, as Angeline exclaimed, "There isn't even anything over that goes over 88!"

Mr. Rohlfing pressed for an interpretation: "So which went higher?"

Abby: "Rounded."

Jen: "I agree, because there's a higher percent" (in the middle and upper bins).

Mr. Rohlfing asked the class which nose cone they would recommend if they were scientists trying to design rockets that could reach a high altitude.

Will said, "The rounded cone. ...If you want something to go higher, then I pick rounded."

Isaac, however, was concerned. He reminded the class that this outcome was inconsistent with their initial expectation: "I don't disagree, but I think it's kind of weird that the, um, the pointed doesn't go as high as the rounded. That doesn't really make sense."

Students then launched into a debate about the mechanisms that might account for this unexpected result. We will not pursue the discussion further here, but it was evident that students were using data distribution as evidence, even though it disconfirmed a cherished belief.

ASSESSMENT OF STUDENT LEARNING

At the completion of instruction, two forms of assessment were administered. The purpose was to probe the extent to which the learning apparently occurring at the class level was also being "picked up" by individual students. First, all students in the class completed a group administered paper-and-pencil assessment consisting of two released items from the 1997 NAEP, designed for students at Grade 4. These items were designed to assess students' capability to compare tables and to read and interpret pictographs. The remaining items were devised by Cobb and colleagues (Cobb, 1999; Cobb et al., in press, McGatha, 2000) to investigate seventh- and eighth-graders' statistical reasoning. These items required students to adjust a given distribution so as to shift the median in a variety of ways.

The whole-class findings were supplemented with individual clinical interviews conducted with 15 of the students. The interview items were devised by us to assess student strategies for reasoning about the probable goal of an experiment given the data, comparing distributions of unequal numbers, reasoning about measurement variability, and comparing two distributions with different shapes, but equal ranges.

Whole Class Assessment

Items for the whole class assessment were prepared in booklets (one item per page) that were group administered, and answers were written individually by all students. Students were provided as much time as they needed to complete the items; all were finished within 25 min.

The first question was adapted from a released item administered in the 1996 NAEP (ID #M061905J). According to NAEP, this item targets students' skills in collecting, organizing, reading, representing, and interpreting data, in particular, interpreting data tables. We showed students Table 1, which displays the votes presumably counted in a classroom election to select a favorite cartoon character (the choices were Batman, Fred Flintstone, and Barney Rubble). As Table 1 shows, votes were taken in each of three classes for the three cartoon characters. Students were told, "Using the information in the chart, Mr. Bell must select one of the characters as the student favorite. Which one should he select, and why?"

We scored student responses with the rubric developed by NAEP for Grade 4, which deemed Batman to be the correct response, as long as that response was justified by either of the following reasons: (1) More students chose it, or (2) Batman was the first choice in one class and second choice in the other two classes. In addition, the NAEP scoring rubric identified three kinds of incorrect responses. The first (labeled Incorrect 1 on Table 2) was choice of Barney Rubble, accompanied by the explanation that referred to the total number of votes received by one or more characters. The second was choice of Batman, but either no explanation is given, or the explanation is inadequate because the student incorrectly computes the number of votes. The remaining incorrect responses—e.g., those that fit neither Incorrect 1 or 2—were categorized into a final group (Incorrect 3).

Table 2 compares the fourth grade 1996 National Performance Results to those obtained in this study. As the table shows, the students in this study easily outperformed the NAEP fourth grade national sample. Of course, comparisons like this are only suggestive, because there are many differences among the populations being compared (e.g., socioeconomic background, stakes of the test) beyond instruction alone that could account for the differences in scores.

The second question selected from the 1992 NAEP test bank (Item number ID #M04900) and administered in 1997 as well, presented three frequency charts, as shown in Figure 3, and asked students to select the chart that might correctly represent the number of pockets worn by 20 students on a particular day in a fictional classroom. According to the NAEP, this item also focuses on collecting, organizing, reading, representing, and interpreting data, in particular, in the form of graphical displays.

Student responses to this item were scored by the NAEP as "Extended," "Satisfactory," "Partial," "Minimal," or "Incorrect." Extended replies were considered to be those in which the student chose the graph labeled B, accompanying the choice with a good explanation why B is the right answer and why A and C are unacceptable.

TABLE 1
Item Adapted From NAEP ID #M061905, "Use Data from a Chart"

Character	*Class 1*	*Class 2*	*Class 3*
Batman	9	14	11
Fred Flintstone	1	9	17
Barney Rubble	22	7	2

Note: In Mr. Bell's classes, the students voted for their favorite cartoon character. Here are the results. Using the information in the chart, Mr. Bell must select one of the characters. Which one should he select, and why? Explain.

TABLE 2
Student Performance on "Use Data from a Chart," NAEP ID #M061905J

Item Score	*1996 NAEP Results*	*This Study*
Correct	32%	86%
Incorrect #3	32%	10%
Incorrect #2	12%	0%
Incorrect #1	21%	4%
No reply	3%	0%

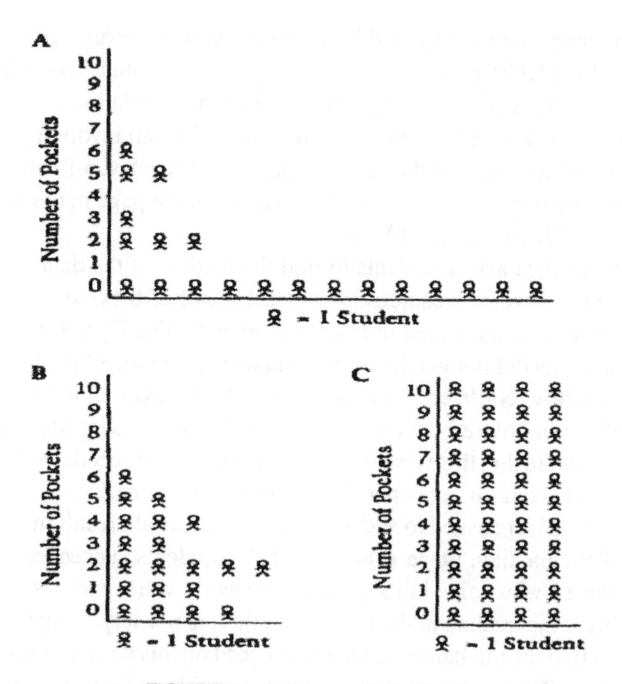

FIGURE 3 Assessment Item, NAEP.

To be scored as extended, the reply must mention both the number of students in the class and the likelihood that most of them have pockets. Responses scored satisfactory were those that included both a correct choice and a good explanation. The explanation might include the fact that B shows 20 students and most have pockets or a good explanation why A and C cannot be right. Partial and minimal replies were those in which the student chose graph B, but did not provide an adequate or relevant explanation. As Table 3 shows, students in this study once again dramatically outperformed the NAEP national sample. Moreover, the results from the national sample suggest that this is a very difficult item for most upper elementary aged students.

TABLE 3
Identify Correct Pictograph, NAEP ID #M04900

	1992 NAEP Results	*This Study*
Extended	3%	67%
Satisfactory	7%	4%
Partial	15%	0%
Minimal	23%	29%
Omitted Item	6%	0%
Incorrect/Off Task	46%	0%

The remaining items completed by all students were developed by Cobb and McClain (Cobb, McClain, & Gravemeiser, in press; McGatha, 2000). These items focused on students' understanding of the median and its relation to other values in a distribution. There were five sub-items in this section, all involving the interpretation and/or manipulation of the identical data set. At each sub-item, the identical data set was presented, arranged from left to right on the page in the following order: 12, 19, 15, 21, 18, 12, 15, 30, 28.

The first sub-item asked students to find the median of the data set. Fifteen of the 22 students answered this question correctly. All of these students first reordered the data and then counted in from both ends to identify the median. The remaining 6 students did not get the correct answer. Of these, 5 failed to reorder the data and 1 partially reordered it. Those who made mistakes all identified the middle value of the unordered data set. Interestingly, one student appeared to be at least partially aware that this was not an effective way to proceed; we found the following note to herself on her paper, "Put in order next time."

Next, students were asked to add a data point that would result in an increase in the value of the median. Once again, 15 of the students succeeded in doing so. Twelve of the 15 who solved this question correctly were students who had correctly identified the median in the first sub-item. It is not surprising that 3 students who failed to find the median were able to succeed on this item, because it is a relatively simple matter to increase the median value without first finding it—perhaps by adding an extreme value.

Sub-item 3 asked students to remove a data point to generate a lower value for the median. This task is arguably more difficult, because students could not simply tack on an extreme value. Sixteen of the students succeeded. Of the students who provided incorrect answers, four removed a data point below the median value, and one failed to eliminate any value.

Sub-item 4 asked students to add a data point to lower the median value of the set. Students found this item more difficult than any of those that preceded it. One conjecture is that students found it confusing to add something that could result in a smaller value (their intuitive ideas about adding presumably are associated with increasing value). Some students even wrote comments on their booklets that suggested they believed that this task was impossible, for example:

Kathy: "It can't be done."
Earle: "Can't be done."
Kevin: "No way I know this."

Altogether, 14 of the students either got the question wrong or provided answers that were ambiguous.

Sub-item 5 asked students to remove a data point that would result in a higher median value. Fourteen of the students provided a correct reply to this question.

Five of the students got all five sub-items correct, and an additional 2 got four correct. One student got all questions wrong, and an additional 3 got most of them wrong. As one might expect, reordering the data set before attempting to answer these items was a strong predictor of success. The students who consistently reordered the data got 80% of these items correct. In contrast, the 6 who failed to reorder the data got only 36% of these items correct. Table 4 summarizes performance on this item.

Individual Interviews

In addition to completing the written assessment tasks, 15 of the students also participated in individual interviews that lasted approximately 30 min each. The interviews were tape recorded and transcribed for later analysis.

Reasoning about experiment. The first interview question was adapted from research previously conducted with other middle school students (Petrosino, 1998). We asked students to make an inference about the most likely purpose for the collection of the data displayed in Table 5. Participants were told that the table displays data collected by other sixth-graders in Nashville, TN. The interviewer

TABLE 4
Number of "Re-orderers" and "Non-re-orderers" Who Got Each Sub-Item Correct

Strategy	*Item 1*	*Item 2*	*Item 3*	*Item 4*	*Item 5*	*Total*
Re-Orderers						
Correct	14	13	12	8	13	60
Incorrect	1	2	3	7	2	15
Non-Re-Orderers						
Correct	1	1	3	2	4	11
Incorrect	5	5	3	4	2	19

continued, "These kids were also shooting off model rockets. Can you look at the table and figure out the purpose of the experiment?"

All 15 students responded that the kids were trying to figure out which went higher, the rounded or pointed cone rockets. Next we asked students to tell us whether the data in the table could support an inference about rocket design. When asked to use a 5-point Likert scale to indicate their confidence that the nose cone design makes a difference, not a single student claimed to be "Very Confident." Ten of the 15 students responded "Pretty Confident" (4 on the 5-point scale), and the remaining 5 insisted that they were "Not Confident at all." When asked why the table does not support a very confident judgment, 8 of the students pointed out the overlap in the distributions, and explained that there were too few data points in which the pointed-cone rockets attained higher altitude than those with rounded cones. Several of these students spontaneously made comparisons to "what we did in class with the medians and spread numbers." When asked what might help make the data more interpretable, students mentioned comparing the medians of the two distributions or eliminating outliers. In addition, 7 students suggested that there should be either "more data" or "more launches".

Interestingly, in previous research conducted by Petrosino (1998), none of the middle school students in that study questioned the interpretability of the data in Table 5. Instead, all readily agreed that the data supported a "very confident" conclusion that the nose cone design affects the height of the rockets. These students had participated in extended discussions about rocket design and flight and about appropriate ways to design experimental comparisons, but had received no instruction about measurement, measurement error, or interpreting and comparing distributions.

Using statistics to compare distributions. The second interview question was another borrowed from Cobb and McClain (McGatha, 2000). Students were asked which of two basketball players, Bob or Deon, should be selected to an All-Star team. Students were to make this decision by comparing the number of

TABLE 5
Data from Nashville Rocket Experiment

		Nose Cone Design		
Group	*Launch Site*	*Rounded*	*Pointed*	*Engine Type*
#1	Playground	210		B4-2
#2	Playground	202		B4-2
#3	Playground	119		B4-2
#4	Playground		133	B4-2
#5	Playground	118		B4-2
#6	Playground		177	B4-2
#7	Playground		123	B4-2

points scored by each player on several games over a season of play. In the data set shown to children, replicated in Table 5, Bob played eight games, compared to only six played by Deon. Moreover, Bob scored more points in total than Deon did. However, Deon achieved a higher average of points per game. The data for each player were presented in unordered fashion on index cards, as shown in Table 6.

Students' initial judgments were recorded, and then the interviewer asked what strategy they had used to come to a decision. Eleven of the 15 of the students initially answered something like, "Bob scored more points" (a result similar to that found in the Cobb study with eighth-graders). In this case, the interviewer followed with a counter-suggestion; for example, "But didn't he play in more games?" Table 7 summarizes the initial and final decisions made and justifications offered by the 15 students. As the Table indicates, there was a notable shift between the initial and final decision; only 3 students continued to advance the "more points" justification. Before (gentle) prompting, most of the students (11 of 15) simply chose Bob, the player who amassed the greatest total number of points. After receiving the invitation to "rethink" their choice, most students backed away from this strategy (two thirds now nominated Deon for the team) and most sought instead to compare either average or median number of points achieved per game (8 students). Four of these students mentioned both average and median. Two students mentioned the range of scores, suggesting that Deon should be selected because his range of scores was higher than Bob's. Two students adopted a novel

TABLE 6
Points Per Game Scored by Two Basketball Players

Player	*Scores*
Bob	21, 16, 23, 21, 20, 17, 16, 22
Deon	24, 18, 21, 25, 22, 28

TABLE 7
Number of Students Selecting Bob or Deon and Justifications for Choice

	Initial Selection	*Final Selection*	*Initial Justification*	*Final Justification*
Bob	11	5		
Deon	4	10		
More Points			11	3
Average or Median Points			3	8
More Games				2
Spread or Range				2

strategy not seen initially, that is, concluding that Bob should be selected because he was a more experienced player (e.g., had played a larger number of games).

Relations between precision and variability. The third question focused on students' understanding of differences in precision of measure that can result from using different measurement instruments. Recall that students had measured the school's flagpole with both a paper measuring device ("height-o-meter") and a plastic altimeter (AltiTrak™). First, students were asked to predict the measurements that might be obtained by a fictional class of 15 children who needed to measure the height of a 100-ft (30.5 m) statue. Students were told that the class measured first with the one instrument and then with the other (order of tools was counterbalanced in the interview). Students were also told that the height of the statue was known by the teacher, but not by the children conducting the measurements. Students had to actually construct the distributions that they might expect to result.

As one would find if in fact the height-o-meter were less precise, the student-generated distributions were notably less variable for the pooled distribution of student-generated AltiTrak™ measurements (SD = 13.9) than for the height-o-meter (SD = 26.7). A matched pairs sign test of the standard deviations of the scores generated by each student indicated that students generated more variable distributions for the height-o-meter than for the AltiTrak™, $p < .01$. Moreover, students' distributions were roughly symmetric as anticipated, with skew values of –.621 and .854 for the pooled distributions of the AltiTrak™ and height-o-meter, respectively. Distributions created by individuals again conformed to this tendency toward symmetry, with only one student in each condition generating a distribution characterized by an absolute skew value greater than 1.0.

Comparing distributions. In the fourth and final interview question, students were asked to compare two distributions that shared some features (same N of subjects, same total sum of values, same means, same ranges), but differed with respect to others (slightly different medians, different spread). The context was Mr. Jones, who owns a tree nursery, and wants to understand whether it matters if he grows his trees in light or dark soil. Table 8 displays the data sets that students were asked to compare. We were most interested in whether or not student responses were influenced by the evident difference in variability. (The variation of trees grown in dark soil was about three times greater than that of trees grown in light soil.) Because one student was called away from the interview before it was completed, only 14 responses were recorded for this item.

Student responses to this item were quite variable. The most common answer (6 of 14 students) was that soil does not matter. Four of these students calculated the average of each distribution, noted that the averages were identical (56.7 ft [17.3 m] min each case), and on this basis, concluded that the distributions did not differ. The remaining ten students also noted similar centers, but their preferences were influenced by the variability of the data. Five students indicated that the dark soil would be

TABLE 8
Heights of Trees Grown in Light and Dark Soil After 3 Months

				Light Soil					
10	*20*	*30*	*40*	*50*	*60*	*70*	*80*	*90*	*100*
19	23		43	51	60	71			109
				52	61	75			
				53	61				
				54	62				
				56	65				
				57	65				
				58					
				Dark Soil					
19	21	32	41		61	75	81	91	109
	22	34	41		61		82	92	
	23							92	
	25							93	

a better choice; these students referred to the fact that the dark soil seemed to produce a greater number of extremely tall trees. Another 3 students preferred light soil, since "most of the trees were in the 50s and 60s." These students pointed out that if you wanted "really" tall trees you should plant them in dark soil, but that there would also be a higher chance that some of the trees would be really small. One student argued, "The soil with the smaller spread is the better soil." Another pointed out that there is "a higher probability of a tree being in the 50s or 60s if it was in the light soil, but the dark soil gives you the best chance of getting a really big tree." This student concluded that there is no difference between the soils, unless one could specify what result was desired, tallest possible tree or highest average height. Table 9 summarizes the conclusions and justifications offered by the 14 students who responded to this item.

Summary of Assessments

In general, student performance on these written and interview tasks was impressive, although we emphasize that very little is known about how students in general should be expected to do on items of this kind. Students in this fourth grade class well outperformed fourth-graders in a national NAEP sample on two items designed to assess interpretation of data from displays (tables and graphs) and performed comparably to seventh and eighth graders in an instructional study of students' statistical reasoning.

In the interviews, we learned that students generated distributions of measures consistent with process of measurement. Their distributions varied with precision of the instrumentation yet retained the symmetry expected when considering error of measure. Lessons learned about distribution were used to reason about experi-

TABLE 9
Student Replies to Interview Question About the Best Soil Type for Growing Trees

Type of Soil	*Strategy*	*N of Students*
Dark soil is better		5 Total
	Extremely tall trees	3
	Median	1
	Average	
	Higher percentage of trees that are average height or taller	
	Average spread	1
Light soil is better		3 Total
	Extreme	
	Median	
	Average	
	Higher percentage of trees that are average height or taller	2
	Average spread	1
Soil type does not matter		6 Total
	Extremely tall trees	1
	Median	1
	Average	4
	Higher percentage of trees that are average height or taller	
	Average spread	

ment: Students were able to induce the plan for an experiment by reviewing the data collected to implement it, and they (unlike previous samples of middle school students) understood that overlapping distributions might imply absence of treatment differences. Students compared distributions of unequal size by referring to measures of center. Finally, when asked to compare two distributions, 10 of the 14 students spontaneously and appropriately took into account the differing variances (one distribution was a normal distribution; the other was bimodal).

DISCUSSION

Over the course of 2 months, fourth-grade students interpreted centers of distributions as representations of true values of attributes, like the heights of flagpoles and the apogee of model rockets. They interpreted the spread or variation of these distributions as representations of processes of measurement, so that increases or decreases in variation came to signal potential transitions in tools, methods, conditions, or measurers (e.g., improving with practice). Hence, measurement served as a context for joint consideration of center and spread, and students came to see that it is difficult to interpret one without the other. This juxtaposition is critical to the development of distributional thinking, in contrast to thinking about statistics as isolated descriptors (Konold & Pollatsek, 2002) We wished to foster a "habit of mind" in which summaries of data were considered jointly, so that ideally children would think of statistics as measures that acted in concert to describe the highlights of a distribution.

Although we could have employed a variety of indicators of variation, our choice of difference scores facilitated a measurer-eye's view of deviation from center as an indicator of precision. The context and practice of measurement helped students appreciate the coordination of distribution of differences and the original measures, so that difference distributions could be more readily appreciated as a rescaling of the original distributions. For example, students expected that most measurements would be "a little over or a little under," and so had similar expectations about differences from the center. Difference scores maintained emphasis on the case-to-aggregate relation, as opposed to other indicators of spread, which tend to obscure this relationship. We believe that this ready coupling of spread with agency rendered the distributions as sensible, so that qualities like symmetry and related aspect of shape came to be understood as anticipated outcomes of measurement processes. This ready correspondence between process and characteristics of distribution is apt to be harder to maintain in other contexts. For example, what is the meaning of the difference between an individual's height and the average height in the population, or transitions in variation over time in such populations? In fact, Porter (1986) suggested that the leap from error to such "natural" variation was historically difficult.

The spadework of measurement distribution set the stage for considering systematic effects (i.e., round vs. pointed nose cones) in light of random effects due to error. Like other scientists, these fourth-grade students had to consider whether an outcome could be considered a treatment effect in light of multiple sources of error variation, such as variation due to trial-to-trial and individual differences. In short, students had to engage in the kind of reasoning at the core of experimental testing, what Mayo (1996) called "arguing from error" (p. 7). It was only after careful consideration and characterization of these sources of variation that the class moved to consider whether their conjecture could be sustained. Here students noticed that the upper tails of a distribution constructed by pooling both effects (launches with round and pointed nose cones) was populated exclusively by the less preferred alternative. Despite ideas about pointed cones "cutting" through the air, consideration of the joint distribution suggested, if anything, the opposite effect: Rounded nose cones might very well be superior. Although studies of scientific reasoning often offer compelling reason to believe that students of this age cannot readily coordinate theory and evidence, our conjecture is that such studies often do not provide students with the tools that professional practice suggests are important. In this case, armed with an emerging repertoire of distribution, students proved capable of coordinating data with conjecture, even though the conclusion that they reached was one that they initially found "insensible." Their puzzling about this resulting "insensibility" set the stage for continued inquiry, again a hallmark of sound scientific practice.

ACKNOWLEDGMENTS

Preparation of this article was funded in part by Grant JMSF–97–10 from the James S. McDonnell Foundation, and from funding by the National Science Foundation (REC–9973004). We all contributed equally to this work. We thank Mark Rohlfing and his students at Country View Elementary School, and the National Science Foundation, which supported this work. The authors thank Paul Cobb, Kay McClain, and Maggie McGatha for their helpful comments during the conceiving and implementation of this study, and Kay McClain for constructive comments on earlier versions of this article.

An earlier version of this article was presented at the annual meeting of the American Educational Research Association in New Orleans, LA, April 2000.

REFERENCES

Gould, S. J. (1996). *Full house*. New York: Three Rivers Press.

Hancock, C., Kaput, J. J., & Goldsmith, L.T. (1992). Authentic inquiry with data: Critical barriers to classroom implementation. *Educational Psychologist, 27,* 337–364.

Konold, C., & Pollatsek, A. (2002). Data analysis as the search for signals in noisy processes. *Journal for Research in Mathematics Education, 33*(4), 259–289.

Lehrer, R., & Schauble, L. (2000). Modeling in mathematics and science. In R. Glaser (Ed.), *Advances in instructional psychology* (Vol. 15, pp. 101–159). Mahwah, NJ: Lawrence Erlbaum Associates, Inc.

Lehrer, R., Schauble, L., Strom, D., & Pligge, M. (2001). Similarity of form and substance: Modeling material kind. In D. Klahr & S. Carver (Eds.), *Cognition and instruction: 25 years of progress* (pp. 39–74). Mahwah, NJ: Lawrence Erlbaum Associates, Inc.

Mayo, D. G. (1996). *Error and the growth of experimental knowledge.* Chicago: University of Chicago Press.

McGatha, M. (2000). Instructional design in the context of classroom-based research: Documenting the learning of a research team as it engaged in a mathematics design experiment. Unpublished doctoral dissertation, Vanderbilt University, TN.

Medin, D. L. (1989). Concepts and conceptual structure. *American Psychologist, 44,* 1469–1481.

National Assessment of Educational Practice. (1992). Identify correct pictograph. NAEP Item M04900. Block: 1992–4M15 No. 10. Washington, DC: Author.

National Assessment of Educational Practice. (1996). Use data from a chart. NAEP Item #M061905J. Block: 1996–4M10 No. 5. Washington, DC: Author.

National Research Council. (1996). *National science education standards.* Washington, DC: National Academy Press.

Petrosino, A. J. (1995). *Mission to Mars: An integrated curriculum* (Tech Rep. SFT–1). Nashville, TN: Vanderbilt University, Learning Technology Center.

Petrosino, A. J. (1998). At-risk children's use of reflection and revision in hands-on experimental activities. *Dissertation Abstracts International-A, 59* (03). (UMI No. AAT 9827617)

Pickering, A. (1995). *The mangle of practice*. Chicago: University of Chicago Press.

Porter, T. M. (1986). *The rise of statistical thinking 1820–1900.* Princeton, NJ: Princeton University Press.

Stigler, S. M. (1986). *The history of statistics: The measurement of uncertainty before 1900.* Cambridge, MA: Harvard University Press.

Weiner, J. (1994). *The beak of the finch.* New York: Vintage Books.

MATHEMATICAL THINKING AND LEARNING, 5(2&3), 157–189

Problem Solving, Modeling, and Local Conceptual Development

Richard Lesh
School of Education
Purdue University

Guershon Harel
Department of Mathematics
University of California at San Diego

The research reported here describes similarities and differences between (a) *modeling cycles* that students typically go through during 60–90 min solutions to a class of problems thast we refer to as *model-eliciting activities*, and (b) *stages of development* that students typically go through during the "natural" development of constructs (conceptual systems, cognitive structures) that cognitive psychologists consider to be relevant to these specific problems. Examples of relevant constructs include those that underlie children's developing ways of thinking about fractions, ratios, rates, proportions, or other elementary, but deep mathematical ideas. Results show that, when problem solvers go through an iterative sequence of testing and revising cycles to develop productive models (or ways of thinking) about a given problem solving situation, and when the conceptual systems that are needed are similar to those that underlie important constructs in the school mathematics curriculum, then these modeling cycles often appear to be local or situated versions of the general stages of development that developmental psychologists and mathematics educators have observed over time periods of several years for the relevant mathematics constructs. Furthermore, the processes that contribute to local conceptual development in model-eliciting activities are similar in many respects to the processes that contribute to general cognitive development.

Applying principles from developmental psychology to problem solving—and vice versa—is a relatively new phenomenon in mathematics education (Lesh &

Requests for reprints should be sent to Richard Lesh, School of Education, LAEB 1440 Room 6130, Purdue University, West Lafayette, IN 47906. E-mail: rlesh@purdue.edu

Zawojewski, 1987; Zawojewski & Lesh, 2003). One implication of this approach is that we expect the mechanisms that contribute to general conceptual development to be useful to help explain students' problem solving processes in individual problem-solving sessions (Lester & Kehle, 2003). Or conversely, mechanisms that are important in local conceptual development sessions should help explain the situated development of students' general reasoning capabilities (Harel & Lesh, 2003). Yet, when problem solving is interpreted as local conceptual development, many currently prevailing views about conceptual development need to be revised substantially (Lesh & Doerr, 2003). For example, it becomes clear that the development of powerful conceptual systems often is a great deal more situated, piecemeal, multidimensional, and unstable than has been suggested by Piaget-inspired ladder-like descriptions of cognitive development (diSessa, 1988). Furthermore, the process of gradually sorting out and refining unstable conceptual systems tends to be quite different than the process of constructing (or assembling) stable conceptual systems (Lesh & Doerr, 1998). In particular, the article by Lesh, Doerr, Carmona, and Hjalmarson (this issue) describes a variety of ways that models and modeling perspectives have required us to move significantly beyond constructivist ways of thinking about mathematics teaching, learning and problem solving. In the meantime, the primary aim of this article is to clarify the nature of local conceptual development—because situated development of powerful, sharable, and reuseable models is a cornerstone concept underlying many of the most significant aspects of our models and modeling perspective.

FOR MODEL-ELICITING ACTIVITIES, PROBLEM-SOLVING PROCESSES USUALLY INVOLVE MULTIPLE MODELING CYCLES

In general, the kind of problem-solving situations that we emphasize are simulations of real life experiences where mathematical thinking is useful in the everyday lives of students, or their friends or families. In particular, many are middle-school versions of the kind of case studies that are emphasized (for both instruction and assessment) in many of our nation's leading graduate schools in fields ranging from aeronautical engineering to business management—where leaders are being groomed for success in the 21st century. We are especially interested in these case studies for children because one of the goals of our research is to clarify the nature of the most important mathematical understandings and abilities that provide foundations for success beyond school in a technology-based age of information (Lesh, Zawojewski, & Carmona, 2003).

A distinguishing characteristic of the preceding case studies for children is that the products that problem solvers produce generally involve much more than simply giving brief answers to well formulated questions. In fact, relevant solution processes usually involve much more than simply getting from givens to goals when the path is blocked (Lester & Kehle, 2003). That is, problem solvers produce conceptual tools that include explicit mathematical models for constructing, describing, or explaining mathematically significant systems. In other words, case studies for children are model-eliciting activities (Lesh, Hoover, Hole, Kelly, & Post, 2000), and the models that problem solvers produce include constructs and conceptual systems that are needed to make sense of the kind of complex systems that are ubiquitous in a technology-based age of information (Lesh, 2001).

> *Models* are conceptual systems that generally tend to be expressed using a variety of interacting representational media, which may involve written symbols, spoken language, computer-based graphics, paper-based diagrams or graphs, or experience-based metaphors. Their purposes are to construct, describe or explain other system(s).
>
> Models include both: (a) *a conceptual system* for describing or explaining the relevant mathematical objects, relations, actions, patterns, and regularities that are attributed to the problem-solving situation; and (b) *accompanying procedures* for generating useful constructions, manipulations, or predictions for achieving clearly recognized goals.
>
> Mathematical models are distinct from other categories of models mainly because they focus on structural characteristics (rather than, for example, physical, biological, or artistic characteristics) of systems they describe.
>
> *Model development* typically involves quantifying, organizing, systematizing, dimensionalizing, coordinatizing, and (in general) mathematizing objects, relations, operations, patterns, or rules that are attributed to the modeled system. Consequently, the development of sufficiently useful models typically requires a series of iterative "modeling cycles" where trial descriptions (constructions, explanations) are tested and revised repeatedly.

DIFFERENCES BETWEEN MODEL-ELICITING ACTIVITIES AND TRADITIONAL WORD PROBLEMS

One of the most significant differences between model-eliciting activities and most traditional textbook word problems is the nature of the problem (beyond difficulties associated with computational skills that such problems generally emphasize), where students must make meaning out of a symbolically described situation. For model-eliciting activities, what is most problematic is that students must make productive symbolic descriptions of meaningful situations. That is, descriptions and explanations (or constructions) are not just relatively insignificant accompaniments to "answers." They are the most critical components of conceptual tools that need to be produced. Therefore, development processes generally involve a series of modeling cycles in which the problem solver's ways of thinking about the "given" information (e.g., givens, goals, and available solution steps) need to be tested and revised iteratively.[1]

Another relevant characteristic that makes model-eliciting activities strikingly different than most traditional textbook word problems is that, in most real life situations where tools need to be developed, it is usually clear: (a) who needs the tool; and (b) why, or for what purpose, the tool is needed. For example, if the goal were to build a boat so that a group of students can cross a river, then the product that the students need to produce is the boat (whose design is unspecified); and, the test of whether the design is adequate is based on the criteria of crossing the river safely. In other words, the purpose provides an "end in view" (Archambault, 1964) that provides a way for problem solvers to assess the adequacy of their current way of thinking about the nature of product design—or strengths and weaknesses of alternative designs (English & Lesh, 2003).

A third distinctive characteristic of model-eliciting activities is that it seldom makes sense to devote much effort toward the development of tools unless the goal is to deal with more than a single isolated situation. That is, tool development is worthwhile mainly when the product is intended to be sharable (with other people), reusable (in other situations), and modifiable (for other purposes). Consequently, some of the most important ways that tools are tested are directly related to these characteristics. In particular, tool development is inherently a social activity (Middleton, Lesh, & Heger, 2003) because the tools need to be sharable with others, as well as being useful to yourself at some later time (or in new situations).

MULTIPLE MODELING CYCLES AND LOCAL CONCEPTUAL DEVELOPMENT

Sometimes, we design model-eliciting activities so that the constructs and conceptual systems that problem solvers need to develop involve the same operational/re-

[1]Traditionally, in mathematics education research and development, problem solving has been defined as "getting from givens to goals when the path is not immediately obvious or it is blocked;" and heuristics have been conceived to be answers to the question: "What can you do when you are stuck?" However, when attention shifts toward model-eliciting activities, in which a series of interpretation cycles are required to produce adequate ways of thinking (about givens and goals), then the essence of problem solving involves finding ways to interpret these situations mathematically. Therefore, heuristics and strategies that tend to be most useful focus on helping students find productive ways to adapt, modify, and refine ideas that they do have, rather than being preoccupied with finding ways to help them be more effective when they are stuck (e.g., in puzzles and games when they have no relevant ideas, or when no substantive constructs appear to be relevant). In general, for problems whose solutions involve multiple modeling cycles, the kinds of heuristics and strategies that are most useful tend to be quite different than those that have been emphasized in traditional problems in which the solutions typically involve only a single interpretation cycle to make sense of the situation.

lational schemes (or cognitive structures) that developmental psychologists have investigated for children's mathematical judgments about the underlying ideas that are involved in the problem—e.g., fractions, ratios, rates, proportions. Consequently, when students significantly extend, revise, or refine their ways of thinking about these constructs during a single relatively brief problem-solving session, we refer to these problem-solving sessions as *local conceptual development sessions*. One reason why this term has proven to be especially appropriate is because the modeling cycles that students go through during 60–90 min problem-solving sessions often appear to be remarkably similar to the stages of development that developmental psychologists have observed, for the same constructs, over time periods of several years (Lesh & Kaput, 1988). Consequently, it is possible to directly observe processes that lead to extensions, refinements, revisions, or adaptations in students' ways of thinking.

Examples of several local conceptual development sessions are described in the next section of this article. All of the problems in the next section involve some type of proportional reasoning (or "scaling up"). So, for each problem, we briefly describe a typical transcript in which the modeling cycles that students go through are similar to stages of development that have been described by Piaget and others concerning children's developing ways of thinking about situations that mathematicians might characterize using the proportion A/B = C/D (Inhelder & Piaget, 1958).

According to Piaget, the most essential characteristic of proportional reasoning is that it involves a relation between two relations (A/B = C/D). That is, the relation between A and B is compared to the relation between C and D (Lesh, Post, & Behr, 1989). Also according to Piaget and other developmental psychologists (Piaget, Inhelder, & Szeminska, 1964), children's proportional reasoning capabilities develop through the following basic stages (Behr, Lesh, Post, & Silver, 1983).

Summary of Piaget's Stages of Development for Proportional Reasoning

Stage #1. Early reasoning about proportions tends to be based on only a salient subset of information that is available; and, quantities or relations that are emphasized tend to be those that are easiest to point to directly. In their primitive responses to Piaget's proportional reasoning activities, students tend to ignore part of the relevant data. For example, in a balance beam task, students may notice only the size of the weights on each arm, but ignore the distance of the weights from the fulcrum. Therefore, reasoning tends to involve qualitative judgments (A < B) about counts, lengths, areas, or other types of quantities that Piaget considered to be based on only concrete operational reasoning (Inhelder & Piaget, 1958).

Stage #2. In some of the earliest situations in which students go beyond reasoning about relations between simple quantities to be able to reason about relations between relations, the relevant relations are considered to be additive (A–B = C–D) rather than multiplicative (A/B = C/D) in nature. That is, the "difference" between A and B is compared to the "difference" between C and D (because these "differences" correspond to quantities that can be perceived directly). For example, if a student is shown a 2 × 3 rectangle and is asked to "enlarge it", a correct response is often given by "doubling" to make a 4 × 6 rectangle. However, if the request is then made to "enlarge it again so that the base will be 9", the same students often draw a 7 × 9 rectangle–adding 3 to both sides of the 4 × 6 rectangle.

Stage #3. In other early situations where students begin to reason about relations between relations, their reasoning is based on pattern recognition and replication. For example, if youngsters are given a simple table of values, like the one below, then they may be able to think about proportions by noticing a pattern which they can then apply to discover an unknown value.

A candy store sells 2 pieces of candy for 8 cents. How much does 6 pieces of candy cost?

- 2 pieces for 8 cents
- 4 pieces for 16 cents
- 6 pieces for 24 cents
- 8 pieces for {?} cents

Note: Some researchers call the preceding strategy a "build up" strategy (e.g., Hart, 1984; Inhelder & Piaget, 1958; Karplus & Peterson, 1970). According to Piaget, this type of reasoning does not necessarily represent "true proportional reasoning" because, to answer such problems correctly, students do not necessarily have to be aware of the reversibility of the relevant operations. That is, if a change is made in one of the four variables in a proportion, the student should be able to compensate by changing one of the remaining variables.

Stage #4. Reasoning is based on true multiplicative proportions. That is, it involves a "second-order" relation between two relations; and, the first-order relations are considered to be multiplicative in nature.

EXAMPLES OF LOCAL CONCEPTUAL DEVELOPMENT

This section briefly describes three local conceptual development sessions for three different model-eliciting activities: The Sears Catalog Problem (Lesh &

Kaput, 1988), The Big Foot Problem (Lesh & Doerr, 2003c), and The Quilt Problem (Lesh & Carmona, 2003). All of these examples involve problems and transcripts that have been analyzed in isolation in other articles that we published in the past. Our purpose here is to focus on trends that become salient only when similarities and differences are examined across several tasks and transcripts. Then, this cross-transcript analysis will set the stage for the analysis of a new transcript in the main results section of this article. More complete descriptions of these transcripts can be found in the previously published articles. Or, complete transcripts can be downloaded from: http://TCCT.soe.purdue.edu/library/

For each transcript that we describe in this section, the solution was generated by a group of three middle school students who were videotaped while they were working in situations that were simulations of situations that might reasonably occur in their everyday lives. Each transcript involved a different group of students. Yet, in each case, the students were enrolled in remedial mathematics classes because of their poor records of performance in past school experiences. Also, in each case, the students were from schools in large urban school districts that served predominantly minority and disadvantaged populations. The students worked on the problems during extended class periods that lasted approximately 90 min; and, to ensure that the students would be familiar with the contexts in which the problems occurred, their teachers typically ask them to read and discuss a relevant article from a math-rich newspaper that our research team produced. One goal of these "warm-up exercises" was to encourage students to engage their "real heads" (not just their "school heads") when they attempted to make sense of the problems and situations. Consequently, teachers usually discussed the newspaper articles on the day before students are expected to begin to work on the related problem. Then, when the problem is introduced, the students tended to waste less time getting oriented to the situations.

The Sears Catalog Problem

To introduce this problem, the students were given four resource documents: (a) a local newspaper from 10 years ago, (b) a 10-year-old "back to school" catalog for a local department store (Sears), (c) a current local newspaper, and (d) a current "back to school" catalog from the same department store. To familiarize students with the context of the problem, a math-rich newspaper article was provided and the accompanying "warm-up" questions were about the "back to school buying power" of $200 at the time that the old newspaper had been published. The goal of the problem was for the students to produce a new newspaper article telling readers (other students) how much money would be needed to have the same "back to school buying power" today that $200 had 10 years earlier.

A Brief Summary of a Sears Catalog Transcript

Interpretation #1. The first interpretation of the problem was based on what Piaget called "*additive pre-proportional reasoning*" (Piaget & Inhelder, 1956). Judgments were based on only a biased subset of relevant information. That is, without much reflection about which items to consider, the group began to calculate "price differences" by subtracting pairs of old and new items that were assumed to be comparable. However, only a few items were considered, which were simply the items that were noticed first; and, no apparent thought was given to how these subtracted differences might allow predictions to be made about future buying power. Later, after calculating several price differences (apparently, in the hope of discovering some sort of "pattern"), the tediousness of computation prompted the students to consider "Why are we doing this anyway?" Then, perhaps in an attempt to answer this question, they began trying to determine a sensible list of items that a "typical" student might want to buy.

A brief (and foolish) transition interpretation. A second brief reconceptualization of the problem was based on the notion that, because 10 years had passed, perhaps items should cost ten times as much! ... Although this "brainstorm" was quickly recognized to be foolish, it appeared to serve several useful functions: (a) It introduced a (primitive) multiplicative way to think about relations between "old prices" and "new prices." (b) It refocused attention on the overall goal of finding a way to determine price increases for a representative group of items—or for any given item.

In general, students' early interpretations were characterized by repeatedly "losing the forest because of the trees"—or vice versa. For example, whereas the first conceptualization lost sight of the overall goal when attention was focused on details (individual subtractive differences), the second conceptualization ignored details when attention was focused on the relation between salaries. For example, the students failed to notice that prices definitely did not increase by a factor of 10 over a 10-year period.

Interpretation #2. The students' second major re-interpretation of the problem was based on what Piaget called "pattern recognition and replication" type of pre-proportional reasoning (Inhelder & Piaget, 1958). That is, the students noticed that the price increases for several items were approximately a factor of 2. So, they guessed that the "new buying power" also should increase by a factor of "about two." The students clearly recognized that this second interpretation showed real promise. However, the new clarity of thought that it introduced also enabled them to notice that (for example) some items actually decreased in price—even though most of the prices increased. So, the group still recognized inadequacies of their current ways of thinking!

Interpretation #3. The students' third major reconceptualization went beyond simple pattern repetition to thinking that involved proportional reasoning much more clearly—and did not involve a simple whole number ratio. It also was based on a more sophisticated way of dealing with the difficulty that not all items increased by the same amount, and that some items actually decreased in price. That is, the students used their calculators to calculate what they referred to as "percent increases" (which were actually just simple multiplicative factors that expressed the relation between an old price and a new price for each item). For example, they found that the sneakers went up by a factor of 2.75; the jeans went up by a factor of 2.25; and, the backpacks went up by a factor of 1.90. Then, these "percent increases" were averaged—even though it was clear that the students were becoming increasingly worried about the facts such as: (a) not all items went up by the same factor, (b) the prices of some items (e.g., calculators) actually decreased, (c) there were disagreements about which items to include or ignore from their list of possible items to buy, and (d) it sometimes was not clear that old and new items were actually comparable.

Interpretation #4. Interpretation #4 explicitly resolved the sampling issue. Also, for the first time, the students actually wrote a (crude) mathematical proportion of the form "A is to B as C is to D"—where the values for A and B were based on sums of prices for a number of "typical" items. Also, whereas interpretation #3 involved finding the average of a collection of "percent increases," interpretation #4 involved finding the "percent increase" for a collection of sums. However, perhaps the most significant insight associated with this new way of thinking was that the students noticed that, after a sufficient number of items had been included in the sum, the factor that they were calling the "percent increase" was not affected much by the addition of more items. In other words, they noticed an intuitive sort of law of large numbers.

The Big Foot Problem

During the class period before the Big Foot Problem was given to the students, the whole class discussed a newspaper article about Tom Brown, the famous tracker, who often works for police to find lost people or to track down criminals (Brown, 1978). Tom Brown is a real person who lives in New Jersey; when he was a young boy, an old Apache "grandfather" taught him to track and live in wild country with no tools or food except what he could make for himself using things he could find. Now, Tom is like the famous detective, Sherlock Holmes. He can find tracks where other people cannot see them; and, just by looking at tracks, he is able to figure out about how big the person is, about how heavy they are, about how fast they are walking or running, and a great many other things. During the first 2 min of the problem-solving session, the observer read the following problem statement to the students. Then, she showed the students a flat box (2" × 2' × 2') that was filled with a 1" layer of dried mud in which a large (size 24 Reebok cross-training shoe) footprint could be seen.

Statement of the Big Footprint Problem

Early this morning, the police discovered that, sometime late last night, some nice people rebuilt the old brick drinking fountain in the park. The mayor would like to thank the people who did it. However, nobody saw who it was. All the police could find were lots of footprints. You have been given a box (Figure 1) showing one of the footprints. The person who made this footprint seems to be very big. Yet, to find this person and his or her friends, it would help if we could figure out how big the person really is. Your job is to make a "HOW TO" TOOL KIT that the police can use to figure out how big people are—just by looking at their footprints.

Your tool kit should work for footprints like the one that is shown here. However, it also should work for other footprints.

A Brief Summary of a Big Foot Transcript

Interpretation #1: Based on qualitative reasoning. For the first 8 min of the session, the students used only global qualitative judgements about the size of footprints for people of different size and sex—or for people wearing different types of shoes. For example, "Wow! This guy's huge. ... You know any girls that big?!! ... Those're Nike's. ... The tread's just like mine."

Interpretation #2: Based on additive reasoning. One student put his foot next to the footprint. Then, he used two fingers to mark the distance between the toe of his shoe and the toe of the footprint. Finally, he moved his hand to imagine

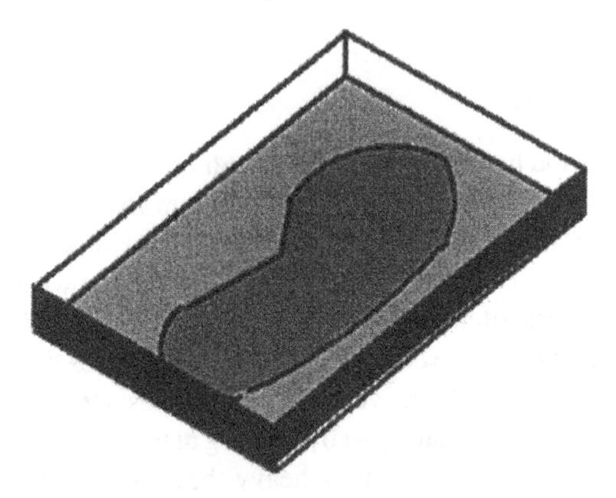

FIGURE 1 Bigfoot print.

moving the distance between his fingers to the top of his head. This allowed him to estimate that the height of the person who made the footprint. However, instead of thinking in terms of multiplicative proportions (A/B = C/D), using this approach, the students were using additive differences. That is, if one footprint is 6" longer than another one, then the heights also were guessed to be 6" different.

> Note: At this point in the session, the students' thinking was quite unstable. For example, nobody noticed that one student's estimate was quite different than another's; and, predictions that did not make sense were simply ignored. Gradually, as predictions become more precise, differences among predictions began to be noticed; and, attention began to focus on answers that did not make sense. Nonetheless, "errors" generally were assumed to result from not doing procedure carefully—rather than from not thinking in productive ways.

Interpretation #3: Based on primitive multiplicative reasoning. Here, reasoning was based on the notion of being "twice as big." That is, if my shoe is twice as big as yours, then I would be predicted to be twice as tall as you."

Interpretation #4: Based on pattern recognition. Here, the students used a kind of concrete graphing approach to focus on trends across a sequence of measurements. That is, they lined up against a wall and used footprint-to-footprint comparisons to make estimates about height-to-height relations, as illustrated in Figure 2. This way of thinking was based on the implicit assumption that the trends

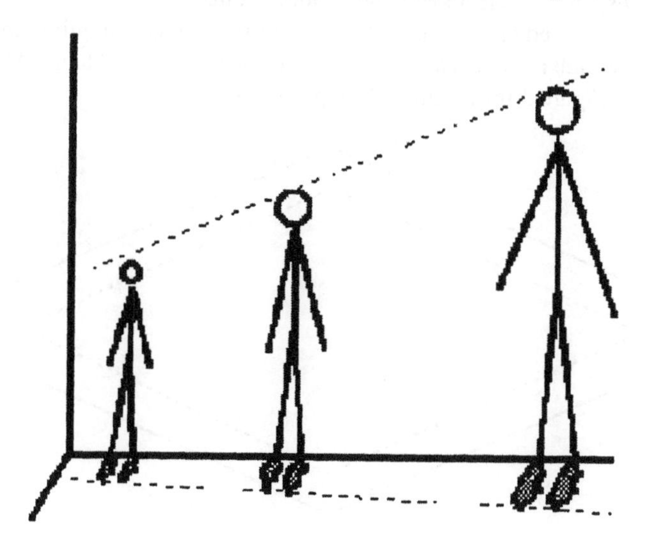

FIGURE 2 Students using a concrete graph.

should be linear—which meant that the relevant relations were unconsciously treated as being multiplicative. The students said:

> Here, try this... Line up at the wall... Put your heels here against the wall.... Ben, stand here. Frank, stand here.... I'll stand here 'cause I'm about the same (size) as Ben. {She points to a point between Ben and Frank that's somewhat closer to Ben}... {pause} ... Now, where should this guy be?—Hmmm. {She sweeps her arm to trace a line passing just in front of their toes.}... {pause} ... Over there, I think.—{long pause}... Ok. So, where's this guy stand? ... About here. {She points to a position where the toes of everyone's shoes would line up in a straight line.}

> Note: At this point in the session, all 3 students were working together to measure heights, and the measurements were getting to be much more precise and accurate than earlier in the session.

Interpretation #5. By the end of the session, the students were being very explicit about comparing footprints-to-height. That is, they estimated that: Height is about six times the size of the footprint. For example, they say: "Everybody's a six footer! (referring to six of their own feet.)"

The Quilt Problem

During the class period before the Quilt Problem was given, the whole class discussed a newspaper article about a local quilting club.

The article showed photographs of several different kinds of quilts; and, it also described how club members made the quilts using patterns and templates like the one shown in Figure 3 for a diamond shaped piece.

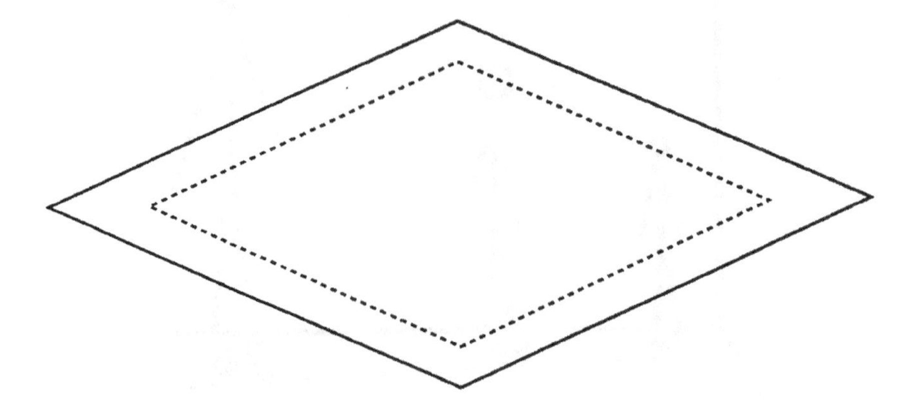

FIGURE 3 A template piece.

FIGURE 4 Example photograph of a quilt.

The problem statement described how quilt club members sometimes had difficulties when they tried to use photographs to make templates that were exactly the right size and shape to make quilts that club members found in books, newspapers, and magazines. So, the job for the students was to write a letter that did two things for the members of the quilting club. (a) First, the letter should describe procedures for making template pieces that were exactly the right size and shape for any quilt whose photograph they might find. (b) Second, the letter should include examples of how to follow their procedures by making templates for each of the pieces of the quilt that is shown in Figure 4.

Note: The quilt that is shown was for a double bed. So, the finished size needed to be approximately 78" by 93".

A Brief Summary of a Quilt Transcript

Interpretation #1a: Focus on a "scaling factor" between the two shapes-as-a-whole—without worrying about the sizes of individual pieces. In this transcript, it is significant to mention that the students whose work is reported had a significant amount of past experience working on other model-eliciting activities that involved scaling-up—or proportional reasoning. In particular, 3 weeks earlier, these same three students had worked on the Big Foot Problem. Therefore, it was natural for them to begin the Quilt Problem by trying to use a process similar to the one that they ended up using earlier. Consequently, they began this session by trying to find a single "scaling factor" that could be used to stretch the picture of a quilt to make it the size of a real 78" by 94" quilt—without worrying about the sizes of individual pieces. Two difficulties arose when they tried to use this approach; the result of these two difficulties led the students to produce results that they did not consider to make sense. First, their ruler-based measurements were not very accurate, and they sometimes were not even correct (e.g., 4 5/8 inches was punched into their calculator as 4.5 inches). Second, the scaling factor that their calculators produced was thought of as a being very messy number (78/4.5 = 17.33333…) that did not seem sensible. (Note: The students themselves were only aware of the second difficulty.)

Interpretation #1b: Scale-up individual pieces within a whole. In attempts to avoid "messy numbers" the students tried to find "scaling-up factors" based on individual pieces of the quilt—rather than being based on the quilt-as-a-whole. (Note: In their earlier attempts, there was no evidence that the whole was viewed as being aggregated from its parts—nor that the same scaling factor should apply uniformly to both the whole and to its parts.) Using this new piece-by-piece approach, new difficulties arose because the scaled-up pieces did not fit together nicely. Thus, the students started trying to measure more carefully (e.g., measuring to the nearest sixteenths rather than eighths of an inch). In spite of these attempts, however, they found that more accurate measurements did not lead to better results. That is, the scaled-up pieces still did not fit together nicely; and, the "scaling-up factors" that they found also continued to be "messy numbers" that they did not consider to make sense.

Interpretation #2: Map from one unit of measure to another within two similar situations. Using interpretation #1b, the students believed that their "errors" (i.e., "messy numbers") must have resulted from measurement errors or from calculation errors. So, they tried to measure more accurately and consistently. In particular, they began to measure all of the pieces by counting sixteenths-of-an-inch markings on their rulers. (Note: The students still had not spoken of what units they were using for their measurements. For example, ¾ of an inch was read simply as ¾.) As a result of measuring in this way, they stumbled on a lucky similarity between the picture and the real quilt. That is, the side of the picture that was supposed to be 93 inches turned out to be 91 (sixteenths-of-an-inch) marks on the ruler; and, similarly, the side that was supposed to be 78 inches turned out to be 76 (sixteenths-of-an-inch markings). So, in both of these instances, they noticed that the difference was 2. This lucky correspondence allowed them to adopt a procedure that essentially translated one sixteenth of an inch (for the picture) to one inch (for the real quilt). This new way of thinking focused on the relations between the units that were used to measure the picture and the quilt—rather than focusing on relations between small and large pieces of the quilt.

Intepretation #3a: Begin to focus on part–part relations within a given quilt (or picture). Partly because the students were becoming increasingly concerned about how the pieces were or were not fitting together, they begin to pay more attention to quantitative relations involving part–part comparisons within a given quilt (or picture). For example, they began to talk about the fact that the sides of the diamonds should be the same as the sides of the small squares—and about the fact that the "big squares in the quilt" should be the sum of several small pieces (i.e., the small squares and triangles). Thus, they sometimes changed "apparent measurements" to force these measurements to give results that they knew should be true (to make the pieces fit together properly).

Interpretation #3b: Explicitly compare wholes and sums-of-parts–and also comparing measurements in two dimensions (height and width). Here, for the first time, the students explicitly stated that when the whole and the parts are scaled-up, the sum of the scaled-parts should be equal to the scaled-up whole. However, no matter now accurately they tried to measure, and no matter how carefully they used their calculators, the sum of the scaled-up parts did not equal the expected scaled-up whole. For example, when they scaled-up the big square (in the picture of the quilt) they got a different result than when they scaled up the small squares and the small triangles.

Interpretation #4: Focus on part–part comparisons based on shapes with shared components (sides). Once the students were convinced that they had found a way to scale up one of the pieces (i.e., the small square), they measured other pieces using this small square as a unit. Instead of applying a single "scaling factor" to each piece of the quilt, they simply scaled up a single piece (e.g., the small square); then they measured all other pieces using this single piece. For example, they recognized that: (a) the little triangles should be half of the little squares, (b) the triangles had two short sides and one longer side, (c) the short sides of the triangles should be the same as the sides of the "kite" shapes. Using this process, they were willing to "adjust" the size of the whole quilt to make the parts fit together properly. In other words, they began to deduce lengths rather than simply measuring lengths; and, they were willing to change what they "saw" (based on direct measurements) to force these measurements to fit what they "understood" (e.g., that the sum of the parts should equal the whole).

> Note: At this point in the session, the students began to act as if they believed they had completed the assigned task. However, when they announced to their teacher that "We're done!" she reminded them that part of their goal was make a template for each of the parts of the quilt shown in the picture. So, the students went back to work.

Interpretation #5a: Part–part comparisons based on angles as well as lengths. Finally, because the students were increasingly precise concerning relations among pieces, and because they were concerned that the "kite shaped pieces" were not turning out quite right, they began to focus on angles as well as the lengths of sides. For example, because the small triangle was half of a small square, they noticed that these triangles must have a "square angle" (i.e., a right angle). Therefore, once they created templates for the triangles and the small squares, they recognized that they could use these templates to figure out the templates for the "kites."

Intepretation #5b: Coordinate all of the Previous Quantitative Relations (Whole–Whole, Part–Whole, and Part–Part—Concerning both Angles and Lengths). Once the students created two comparable pieces for the small quilt and the real quilt, all of their other quantitative comparisons focused on part–whole and part–part comparisons—including relations involving both angles and shared sides and lengths. In particular, because they now had a much more clear understanding that the sum of the parts should be equal to the whole (big square), they checked carefully to make sure that: (a) the templates for all of the pieces fit together to make a whole (i.e., big square), and (b) the big squares plus their borders fit together to make a quilt that was the correct size. To make all of the pieces fit together properly, the students went through several construct>test>revise cycles.

Interpretation #6: A two-stage process focusing on symmetry relations—using the big square (rather than the little square) as the basic unit to be scaled-up. Here, for the first time, the sizes and shapes of the smaller pieces were derived from the shape of the large square (and the quilt-as-a-whole) rather than deriving the shape of the large square from the sizes and shapes of the smaller pieces. That is, the students final way of thinking about the quilt problem used a two-step process: (a) First, determine the size of big squares, the borders, the thin strips, and the quilt-as-a-whole. (B) Then, use paper folding (as illustrated in Figures 5–8) to determine the sizes and shapes of the small pieces within the large squares.

1. Start with a 12″ × 12″ paper square, and fold it along the diagonals (Figure 5).
2. Fold the paper squares vertically and horizontally (Figure 6).

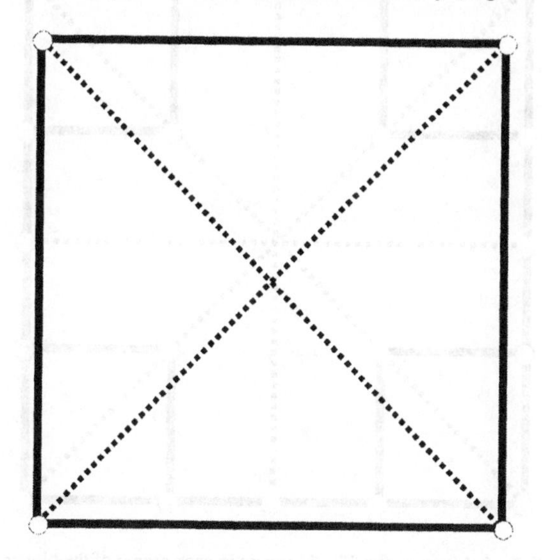

FIGURE 5 1. Start with a 12″ × 12″ paper square, and fold it along the diagonals.

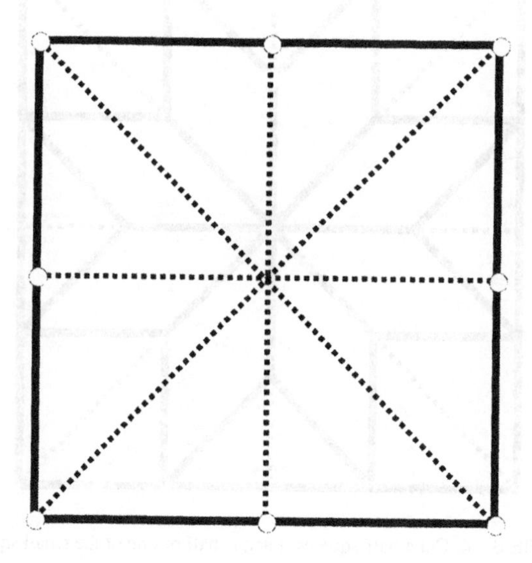

FIGURE 6 2. Fold the paper squares vertically and horizontally.

3. Measure one 3″ × 3″ square in each corner of the bigger square (Figure 7).
4. Cut 4 half-squares. Each is half of one of the small squares (Figure 8)

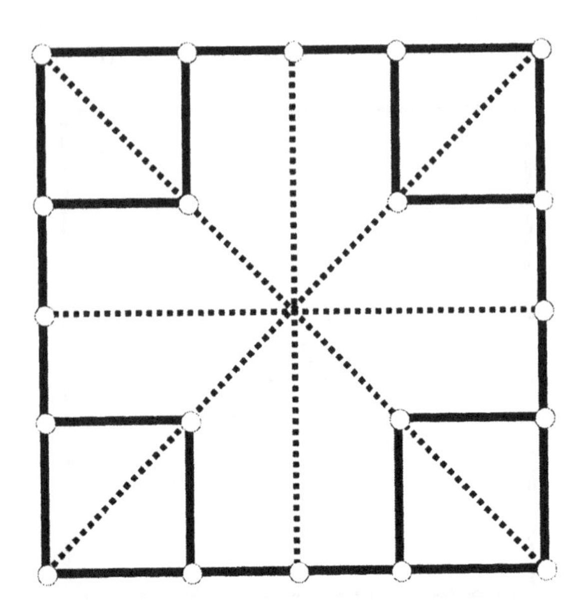

FIGURE 7 3. Measure one 3″ × 3″ square in each corner of the bigger square.

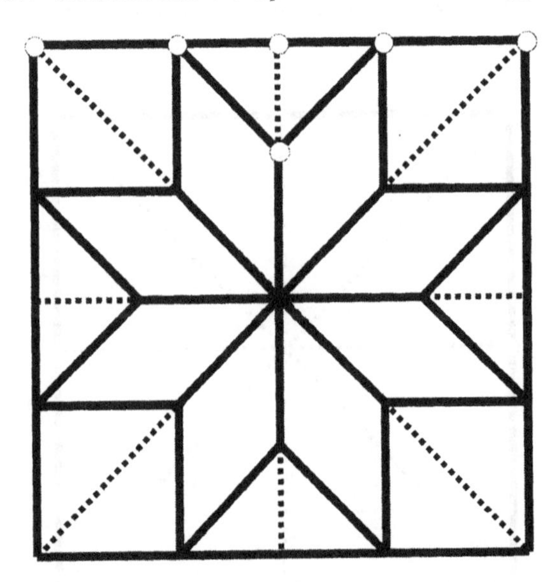

FIGURE 8 4. Cut 4 half-squares. Each is half of one of the small squares.

CONCLUSIONS RELATED TO LOCAL CONCEPTUAL DEVELOPMENT

All three transcripts described in the preceding section illustrate the fact that, when students are able to make sense of situations based on extensions of their own personal knowledge and experiences, and when they express their ways of thinking in forms that they themselves are able test and refine repeatedly, they often invent (or significantly modify, extend, or revise) mathematical constructs that are considerably more sophisticated than those that they seemed unable to comprehend during past histories of failure in situations centered around traditional textbooks, tests, and teaching. In particular, our research has shown consistently that youngsters who are among the least advantaged often invent more powerful ideas than anybody has ever dared to try to teach to them.

The transcript for the Sears Catalog Problem (Lesh & Kaput, 1988) was the first one where we began to recognize the striking similarities that often exist between the modeling cycles that students go through during model-eliciting activities and the stages of development that developmental psychologists have observed for the relevant mathematical ideas (Lesh & Kaput, 1988). This insight first became clear in the videotape analysis laboratory for a series of projects that are known collectively as The Rational Number Project (Post, Behr, & Lesh, 1982). One table in the lab contained drafts of a paper that we were writing about Piaget's description of the development of children's proportional reasoning concepts (Lesh, Post, & Behr, 1989); and, a second table contained drafts of a paper that we were writing about modeling cycles that problem solvers typically go through during solutions to the Sears Catalog Problem. What we noticed was that the stages described on one table were nearly identical to the modeling cycles described on the second table.

The preceding observation also applies to the Big Foot Transcript and the Quilt Transcript reported in the preceding section. One reason why this insight has proved to be significant is because, when students move through a series of Piaget-style stages during a single problem-solving session, it is possible to go beyond investigating states of development to directly observe processes and paths that lead from one state to another. Also, because model development can be induced by putting students in situations where they express their thinking in ways that they themselves can test and revise, and because students themselves determine directions for improvement, it is possible to go beyond investigating typical development in natural environments to also examine induced development within carefully controlled and mathematically enriched environments (Lesh et al., 2000). In other words, it is possible to go beyond investigating the development of general cognitive structures that emerge during periods of major cognitive reorganizations (when children are approximately 6–7 or at 11–12 years of age) to focus on transition periods of development (before/between/following the ages of 7 and

12) for powerful and highly specialized mathematical constructs that seldom develop beyond primitive stages unless artificially enriched mathematical experiences are provided.

Because all three of the transcripts described in the preceding section involved some type of proportional reasoning (or "scaling-up"), an examination of similarities and differences across all three transcripts revealed several important facts that are not apparent in analyses of any single transcript. For example, a student's apparent level of development not only changes significantly during the course of solving each individual model-eliciting activity, it also changes significantly as the student moves from one activity to others that are structurally similar. Thus, just because a student's way of thinking about one problem has moved from stage N to stage N + n ($n = 1,2,3,\ldots$), this does not guarantee that he or she will immediately function at this highest level when beginning to work on another problem that's structurally similar. For example, initial conceptualizations of the second problem may revert back to thinking at level N or N–1. This fact is especially apparent in the complete transcripts that are available to be downloaded for the Big Foot Problem. This is because transcripts are given in which a single group of students worked on a series of structurally similar model-eliciting activities.

The preceding observation could be interpreted as an instance of Vygotsky's concept of zones of proximal development (Vygotsky, 1978). That is, at any given moment in any given situation, a student may have accessible a range of conceptual systems that could be engaged; and, which one does get engaged depends on a variety of factors that might include: guidance from an adult or peers, which among a variety of representational systems (or other conceptual tools) happen to be employed.

An implication of the preceding observations is that, when students begin to work on a model-eliciting activity, relevant constructs almost always should be expected to be at some intermediate stage of development. That is, they be expected to be "completely undeveloped" (so that they need to be constructed beginning with a "blank slate") nor should they be expected to be "completely mastered." In general, the challenge for students is to extend, revise, reorganize, refine, modify, or adapt constructs that they do have—not simply to assemble constructs that are completely new.

Another related observation is that, even though the Sears Catalog Problem, the Big Foot Problem, and the Quilt Problem all involve some type of proportional reasoning, the situations also are quite different. For example, the Sears Catalog Problem gives an overwhelming amount of information that must be filtered, weighted, and organized in some way; whereas, the Big Foot Problem gives very little information directly; most needs to be generated. Consequently, for these and other reasons, each of the three problems involves not only ideas about "scaling-up," but their solutions also draw on some other fundamentally different categories of ideas. For example, in the Sears Catalog Transcript, the students needed to develop some very important thinking about sampling and variability. Or, in the Big Foot Transcript, they introduced some important thinking about graphing, as well as about sampling and variability. Similarly, for the Quilt Transcript, the students developed a series of progressively more sophisticated ways of thinking about relevant units of measure. Consequently, the modeling cycles that students went through for the Quilt Problem looked more like the stages of development that Piaget and others have described for measurement concepts—rather than stages that have been described for proportional reasoning.

An important point to notice about the preceding observations is that it is seldom possible for problem solvers to give an adequate description of a nontrivial "real life" situation using only ideas from a single isolated topic area or discipline. Most solutions inherently involve integrating ideas from a variety of topic areas and disciplines. Consequently, a large part of what it means to understand ideas in any given topic area or discipline involves establishing relations with ideas in other topic areas and disciplines.

To help clarify the meaning of the preceding observations, the final sections of this article describe a transcript for a fourth model-eliciting activity that focuses on ideas in another topic area that is closely related to ideas about proportional reasoning and scaling-up. The topic area is projective geometry; the problem is called the Shadows Problem; and, the main point is that Piagetian stages of development are not the only kind that are useful for describing modeling cycles during model-eliciting activities. For example, Van Hiele's stages of geometric understanding also may appear to be similar to modeling cycles that emerge in some model-eliciting activities.

What's similar about the theories of both Piaget and Van Hiele is that both considered mathematical thinking to be about seeing (or interpreting) situations at least as much as it is about doing algorithmic procedures. Also, both focus on holistic properties of conceptual systems that students use to interpret their experiences. That is, they focus on properties of systems-as-a-whole that are not simply derived from properties of their constituent elements. For example, Piaget's conservation tasks all focus on children's understanding of properties that are invariant under some system of transformation. Therefore, by exhibiting an understanding of invariance with respect to a system, students automatically demonstrate that they are using the relevant system-as-a-whole to interpret their experiences (Lesh & Carmona, 2002).

A BRIEF DESCRIPTION OF VAN HIELE'S STAGES OF DEVELOPMENT

The following stages provide one way to characterize Van Hiele's description of the development of students' geometric understandings (Hoffer, 1983; Van Hiele, 1986)

- First, geometric shapes are considered to be alike or different based on global characteristics of the shapes-as-a-whole. Thus, a rectangle (⎕) might be recognized as being the same as the square (□) simply because "it looks square" to a child. Whereas, a diamond (◊) might not be recognized as being a square—simply because "it doesn't look square." ... Similarly, if a young child is shown the shadow of a tilted circle (so the shadow looks like an oval), and if she is asked to "draw the shape of the shadow that you see," she often will refuse to draw anything that does not look like a circle.
- Second, geometric shapes are considered to be alike or different based on shared properties. So, a square is considered to be "a shape that has four equal sides" (or four equal angles). Yet, a shape that has certain properties is not the same thing as a shape that is defined in terms of these properties. If S is the shape, and $P_1, P_2, \ldots P_n$ are the properties, then the distinction we're referring to is the difference between the statement "S implies $P_1, P_2, \ldots P_n$" and the statement "$P_1 + P_2 + \ldots + P_n S$ implies S."
- Third, geometric shapes are defined in terms of their properties, and by relations among their properties; additional properties can be deduced that might not be immediately obvious by inspection.
- Fourth, geometric shapes are enhanced using auxiliary elements or properties, or by embedding them within larger shapes or systems.

Another Example of Local Conceptual Development

To introduce students to the Shadows Problem, they usually are asked to read and discuss two math-rich newspaper articles about a science fair where the exhibits will focus on the theme of perception and illusion. One article describes similarities between (1) photographs, (2) shadows, and (3) slices of a cone or a pyramid (see Figure 9).

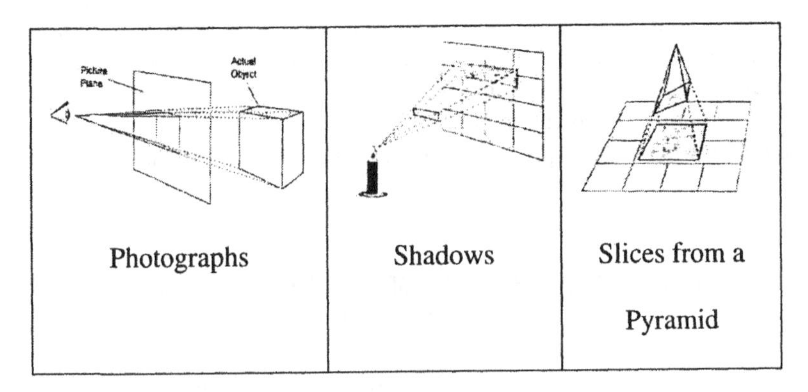

FIGURE 9 Sight is like light and cutty shapes.

The second newspaper article described how some students were able to create illusions using shadows, photographs, or "peek holes" in distorted rooms. For example, Figure 10 shows how a peek-hole in a large box can be used to make two identical toy dolls appear to be completely different in size.

Similarly, Figure 11 shows how a peek-hole in a large box can be used to make a three dimensional shape look like a two dimensional triangle.

The problem that was given to the students told them about a group of students who were hoping to set up an exhibit in which a point source of light would be used, and nonsquare shapes (like the ones shown in Figure 12) would be used to make square shadows. The problem statement asked the students to write a brief two-page letter to the client describing: (a) which of the following figures can be used to make a square shadow, and (b) exactly how should the figure be held, relative to the light and the wall, to make a square shadow. The students were given the

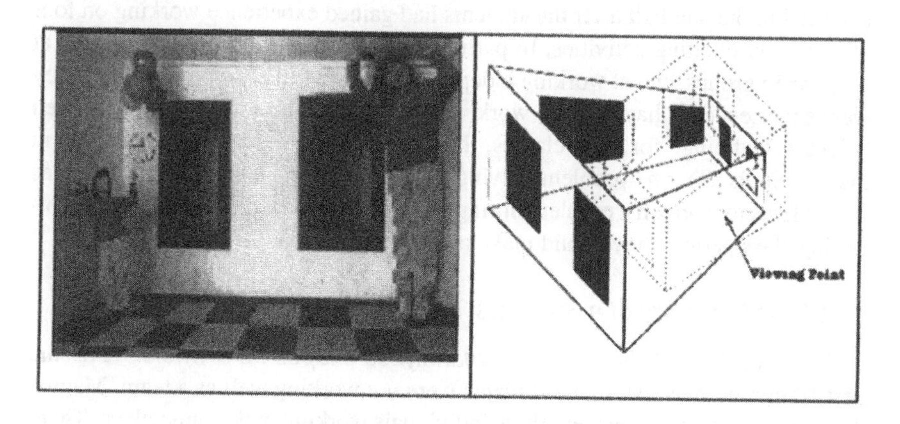

FIGURE 10 Perceptical illusion.

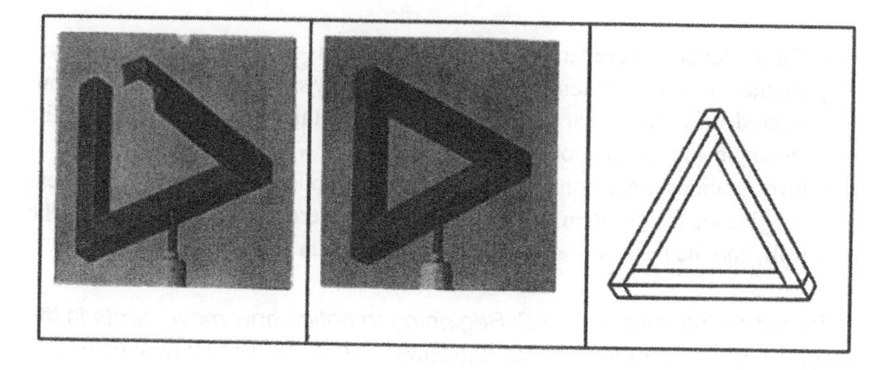

FIGURE 11 Another perceptical illusion.

FIGURE 12 Carboard shapes for making shadows.

six cardboard shapes shown in Figure 12. For a light source, they also were given a small penlight flashlight with the reflector painted black so that it will come close to providing a "point source" of light.

A complete transcript for one group of students is given at the following web site: http://TCCT.soe.purdue.edu/library/. Again, the students (Al, Bev, & Candy) whose solution to the shadows problem is described were inner city Afro-American students who were in a remedial math course for eighth graders. The session occurred in the late Fall after the students had gained experience working on four other model-eliciting activities. In particular, these students were experienced at working in groups, and at working independently for a full class period. Also, they were experienced at having their work videotaped; and, because their school used "block scheduling" for their classes, the class periods were twice as long as in many schools. So, on "problem-solving days" the typical classroom routine was for students to work on a problem during one full class period; then, during the following class period, they would make presentations about their work.

A Brief Summary of a Shadows Transcript

Interpretation #1: Focusing on unanalyzed shapes-as-a-whole. For the first 10 min of the session, the students were not working well as a team. Mainly, they functioned as if they were three individuals working in the same place. There was little collaboration; and, little communication was occurring to coordinate their efforts.

- The students appeared to be thinking of shapes as-a-whole instead of analyzing them in terms of their properties. So, the only shapes that were being considered were those that seemed to "look something like squares" (i.e., the rectangle and the diamond).
- Investigations were constrained by several implicit assumptions that were not given in the problem. For example, shapes were never tilted relative to the wall, and the light was always held perpendicular to the wall.

Transition Interpretation # 2: Beginning to notice how movements in the shapes produce changes in the shadows. Here, for the first time in the session, the students explicitly began to tilt and turn the cardboard shapes and to carefully observe corresponding changes in the shadows that were produced. However, even though the students were beginning to pay attention to ways the shadows change when the shapes were manipulated, they were not exploring transformations in any systematic way.

Interpretation #3: Implicitly thinking of similarities between vision and shadows. Without making their thinking explicit, the students were beginning to reason by analogy—implicitly using "the geometry of vision" to make predictions about "the geometry of shadows." However, the students did not follow up on this analogy; they continued to think in terms of unanalyzed shapes-as-a-whole. For example, just like the early stages of reasoning identified by Van Hiele (1959), Bev refused to recognize a diamond as being a square—simply because its sides were not vertical and horizontal. In general, up to this point in the session, the students seemed to be making the implicit assumption that a shape will only make a square shadow if it already "looks something like a square" by having either: (a) all of its sides equal, or (b) all of its angles equal.

Interpretation #4: Focusing on details as shapes are transformed. Here, for the first time, the students clearly begin to investigate how continuous transformations in one figure lead to other kinds of continuous transformations in the shadow. For example, Al held up the rectangle and turned it very slowly in the light—while watching the shadow change. Unlike earlier in the session, the three students are beginning to work together well. For example, one held the light, one moved the shape, and one checked the shadow—while each was paying attention to what the others did and said.

Interpretation #5: Focusing on relations among properties of shapes. Whereas these students' early interpretations were based on unanalyzed shapes-as-a-whole, just as in the early stages of conceptual development described by Van Hiele, they now began focusing on properties of shapes or on shapes whose properties are recognized. Also, they explicitly began to investigate changes in properties as different shapes were transformed.

Interpretation #6: Taking into consideration the source and direction of the light. Up until this point in the session, the students had not consciously manipulated the direction the light was shining. They had simply shone the light perpendicular to the wall and moved the cardboard shapes. Now, they began to shine the light at an angle to the wall. This was partly done because they sometimes moved the shape and light source so that part of the shadows went beyond the side of the wall that they were using as a projection screen. The students began to hold a shape fixed and move the light source. The students were also explicitly investigating how the shadow changes when the light source moves. So, they were thinking about relations that involve changes in the light source, the cardboard shapes, and the shadow. As they explored the preceding relations, they were becoming conscious of increasing numbers of details about shapes, positions, movements, and shadows. For example, for the first time, the students noticed that, when the shadows get too big, the edges of the shadow get fuzzy.

Interpretation #7A: Focusing on transformations of sides of shapes, rather than shapes-as-a-whole. Here, for the first time, the students explicitly focused on transforming a single side of a shape. They were not simply transforming the shapes-as-a-whole. For example, Bev focused on the longest side of the trapezoid, and she noticed that she could make the shadow of this side as long (or as short) as she wanted—by tilting it or by moving it closer to the light source. Also, for the first time, the students explicitly investigated relations involving pairs of sides—rather than simply focusing on the size of individual sides, or general shapes-as-a-whole.

Interpretation #7B: Focusing on transformations of sides and angles, rather than on shapes-as-a-whole. This was the first time in the session that the students were explicit about paying attention to variations in angles—apart from the contributions that angles made to the general shapes-as-a-whole.

Interpretation #8A: Focusing on complex transformations consisting of a series of simpler transformations. Here, the students tried to produce the shadows they wanted using a series of simpler transformations—each focusing on a single attribute of the final result (rather than simply transforming shapes-as-a-whole). This is the first time that the students tried to hold one characteristic constant while varying other characteristics. For example, Al made two of the adjacent sides of the rectangle "almost the same size." Then, he tried to keep these two sides the same while tilting the rectangle to make the size of the interior angle look "more square" (i.e., 90 degrees). (Note: While doing this, he did not pay any attention to the other sides or angles.)

Interpretation #8B: A shape is defined by specific properties. At this point in the session, the students were being very careful about inspecting all of the properties of the shadow—not just one or two properties that they happened to notice. Also, the students were beginning to go beyond thinking that "if it's a square then it has properties A, B, and C" and they started thinking that "if it has properties A, B, and C, then it's a square." That is, a square was considered to be defined by its properties—rather than simply being a shape that has these properties. For example, for the first time, for a given shape, the students systematically paid attention to all four sides and all four angles at the same time.

Interpretation #9: Going beyond "looking square" to really "being square". This was the first time that the students seemed to recognize the need for a "proof" (or a sensible explanation) that the shadow went beyond simply "looking square" to actually "being square" (by having the necessary properties that define squareness).

The letter below is a digital simulation of the one that was written by Al, Bev, and Candy with formatting and editing help from their teacher. While Bev was busy writing, Candy and Al talked (mainly "off task")—but continued to play with the shapes, light, and shadows. In addition to using the cardboard shapes, they also used their hands to make more "animal faces" and other funny shadows. Then, Al tore the diamond in half to make two equilateral triangles; and, he asked Candy "How many different triangles do you think we can make?" They concluded that: (a) "We can make any triangle we want." (b) "We can make the angles as big or as little as we want." (c) "We can make the sides as big or as little as we want." They demonstrated Point #1 using a broken pencil that the students had used throughout the session.

Dear Students,

You can make square shadows with these shapes (see Figure 13).
Shapes like this one *do not work* (see Figure 14).

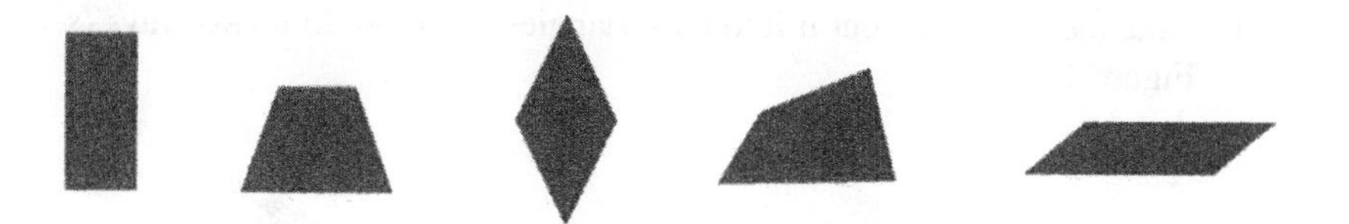

FIGURE 13 Cardboard shapes for making shadows.

FIGURE 14 Shapes like this one don't work.

Follow these three steps.

1. Make one corner square.
2. Make the two sides the same.
3. Go on around and fix each of the sides and corners.

It works. We showed it.
Bev, Candy, and Al

On the next day after the students had finished writing their letters to their client (who was assumed to be some students in another school), they made brief 5-min presentations to their whole class about how they solved the Shadows Problem. When Bev, Cindy, and Al gave their presentation, Bev read the letter, and Candy and Al demonstrated with the light and cardboard shapes. When the teacher asked them to go beyond explaining how their method worked to also show why it worked, Al explained by tearing one of the quadrilaterals in half to make two triangles. Then, he showed how to make "half of a square" with this triangle. Finally, he drew a sketch, like the one shown in Figure 15, to show: (a) how any of the convex quadrilaterals can be cut be cut into two triangles, and (b) how any of these triangles could be used to make "half of a square shadow." After all the students in the class had made presentations about their work, Al, Bev, and Candy revised their letter to include the four steps shown below.

1. Take the shape and cut it into two triangles. You can do it two ways (see Figure 15).

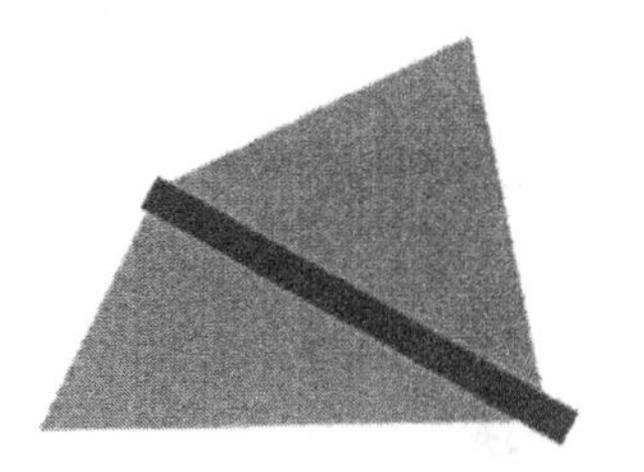
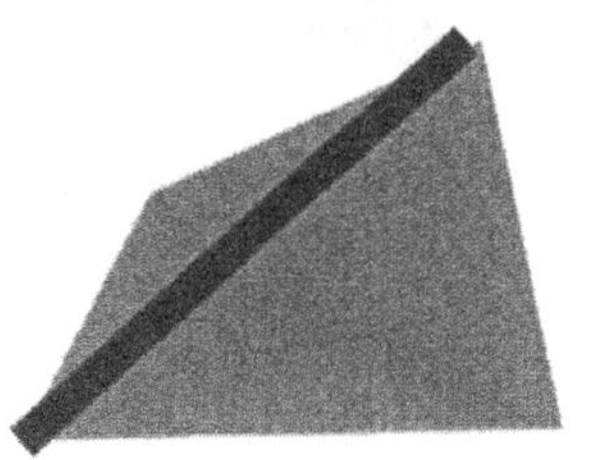

FIGURE 15 1. Take the shape and cut it into two triangles. You can do it two ways.

FIGURE 16 2. Use straws to make kite pieces.

2. Use straws to make kite pieces (see Figure 16).
3. Use only the kite pieces and tilt them to make a square shadow. It is easy (see Figure 17).
4. Put the paper back on the kite and the shadow will look square too (see Figure 18).

Interpretation #10: Using auxiliary lines and shapes, implicitly introducing movements in three dimensions, and providing an informal proof. Whereas the students' first letter described a method that required a lot of tinkering to make it work, their second letter constituted an actual informal proof that "a square shadow can be made using any convex quadrilateral." The procedure they specified used the kind of auxiliary lines (and triangles) shown in Figures 15–18. Also, the students implicitly described movements in three dimensions. That is, they slid the shape in one direction, they turned it in a second direction, and finally, they turned it in a third direction.

FIGURE 17 3. Use only the kite pieces and tilt them to make a square shadow.

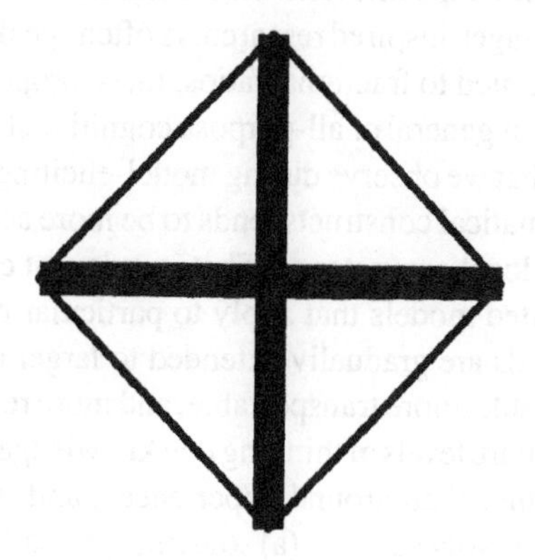

FIGURE 18 4. Put the paper back on the kite and the shadow will look square too.

CONCLUSIONS

For each of the four transcripts described in this article, our analyses focused on the modeling cycles that students went through during a 90-min problem-solving session. In other publications, and for other purposes, we have focused on other ways of thinking about students' behaviors. For example, Zawojewski and Lesh (2003) analyzed similar transcripts by focusing on the problem-solving strategies and heuristics that the students employed; Zawojewski, Lesh, and English (2003) focused on changes in group functioning throughout solution processes; Middleton, Lesh, and Heger (2002) focused on social and affective factors that impact students' behaviors; Harel and Lesh (2002) focused on proof strategies for the same Shadows Transcript that we described here. Readers may wish to use still other perspectives to analyze the transcripts we have described. Complete transcripts can be downloaded from the following web site: http://TCCT.soe.purdue.edu/library/

For the purposes of this article, the first point that we want to emphasize is that, when a model-eliciting activity requires students to develop a conceptual tool that involves a construct or conceptual system whose stages of development have been investigated by developmental psychologists or mathematics educators, then the task-specific modeling cycles often bear striking similarities to corresponding general stages in the development. For example, both Piaget's and Van Hiele's theories of conceptual development can be used to help understand and explain many of the behaviors students exhibit during the process of developing conceptual tools needed in model-eliciting activities. Consequently, it is reasonable to expect that processes that contribute to general cognitive development also should contribute to progress through the modeling cycles for model-eliciting problems.

A second related cluster of implications is related to the fact that, if we examine a student's performances across a series of related activities, it is clear that his or her apparent stage of development often varies considerably across tasks as well as across modeling cycles for a specific task. This is significant for a variety of reasons. For example, whereas Piaget-inspired researchers often speak of mathematical constructs (such as those related to fractions, ratios, rates, proportions) as if they were specific manifestations of general or all-purpose cognitive structures, the local conceptual developments that we observe during model-eliciting activities suggest that the evolution of mathematical constructs tends to be more accurately characterized as gradual increases in local competence. That is, relevant conceptual systems are developed first as situated models that apply to particular problem-solving situations. Then, these models are gradually extended to larger classes of problems as they become more sharable, more transportable, and more reusable. In other words, it is only at relatively mature levels of thinking that knowledge begins to be organized around abstractions rather than around experiences; and, these mature levels of thinking are not likely to evolve unless: (a) students are challenged to develop models and conceptual tools that are sharable, reuseable, and transportable; (b) students are introduced to powerful representation systems for expressing relevant constructs; and (c) students are encouraged to go beyond thinking with these constructs to also think about them.

On the other hand, if these latter functions are emphasized, then significant generalizations often can be expected from one task to other similar tasks. Therefore, when we speak about local conceptual development, we are not suggesting that all learning is task specific—nor that there is no generalization or transfer of learning. In fact, by emphasizing the development of conceptual tools that are sharable, reusable, and transportable, it is much more likely that the relevant constructs and conceptual systems will be useful beyond the situations in which they were created. For example, consider the simple case when a problem is solved using spreadsheet-generated graphs. As anybody knows who has used such tools, some spreadsheets are easier than others to adapt to new data, new purposes, or new situations. Yet, even if the spreadsheet is designed to facilitate transfer, the students who created the spreadsheet may or may not be able to use the tool in a new situation problem-solving situation.

A final point becomes apparent when we examine details within transcripts where this brief report was only able to describe general stages in development. That is, contrary to the ladder-like descriptions of development that stages suggest, details in the transcripts show that students often switched back and forth among a variety of different ways of thinking about a problem-solving situation—usually without noticing that these conceptual shifts had been made. In fact, even within the thinking of individual students, communities of constructs and conceptual systems often were competing to dominate the interpretations that would be emphasized at any given moment. Thus, at any time during a series of modeling cycles, "communities of mind" were apparent not only within the thinking of teams, but also within the thinking of individuals within the teams. So, solution processes often involved gradually differentiating and/or integrating (or dismissing, or refining) alternative ways of thinking—rather than simply progressing along ladder-like sequences. In other words, regardless of what developmental theory is used to explain students' performances during model-eliciting activities, it tends to be far too simplistic to refer to a given child as being at stage N in some developmental sequence. Communities of relevant concepts tend to be available at any given moment; most of these constructs are at intermediate stages of development, and apparent levels of development vary across tasks as well as across time within a given task.

REFERENCES

Archambault, R. D. (Ed.) (1964). *John Dewey on education: Selected writings*. Chicago: University of Chicago Press.

Behr, M., Lesh, R., Post, T., & Silver, E. (1983). Rational-number concepts. In R. Lesh & M. Landau (Eds.), *Acquisition of mathematics concepts & processes* (pp. 91–126). New York: Academic Press.

Brown, T. Jr., & Watkins, W. J. (1978). *The tracker.* New York: Berkley Books.

DiSessa, A. A. (1988). Knowledge in pieces. In G. Forman & P. B. Pufall (Eds.), *Constructivism in the computer age* (pp. 49–70). Hillsdale, NJ: Lawrence Erlbaum Associates, Inc.

Doerr, H. M., & Lesh, R. (2003). A modeling perspective on teacher development. In R. Lesh & H. M. Doerr (Eds.), *Beyond constructivism: Models and modeling perspectives on mathematics teaching, learning, and problem solving.* Mahwah, NJ: Lawrence Erlbaum Associates, Inc.

English, L., & Lesh, R. (2003). Ends-in-view problems. In R. Lesh & H. M. Doerr (Eds.), *Beyond constructivism: Models and modeling perspectives on mathematics teaching, learning, and problem solving.* Mahwah, NJ: Lawrence Erlbaum Associates, Inc.

Harel, G., & R. Lesh (2003). Local conceptual development of proof schemes in a cooperative learning setting. In R. Lesh & H. M. Doerr (Eds.), *Beyond constructivism: Models and modeling perspectives on mathematics teaching, learning, and problem solving.* Mahwah, NJ: Lawrence Erlbaum Associates, Inc.

Hart, K. (1984). *Ratio: Children's strategies and errors. A report of the strategies and errors in secondary mathematics project.* London: The NFER-Nelson Publishing Company, Ltd.

Hoffer, A. (1983). Van Hiele-based research. In R. Lesh & M. Landau (Eds.), *Acquisition of mathematics concepts and processes* (pp. 205–227). New York: Academic Press.

Inhelder, B., & Piaget, J. (1958). *The growth of logical thinking from childhood to adolescence.* New York: Basic Books.

Karplus, R., & Peterson, R. (1970). *Acquisition of mathematics concepts and processes* (R. Lesh & M. Landau, Eds., pp. 7–44). New York: Academic Press.

Kehle, P. E., & Lester, F., Jr. (2003). A semiotic look at modeling behavior from problem solving to modeling: The evolution of thinking about research on complex mathematical activity. In R. Lesh & H. M. Doerr (Eds.), *Beyond constructivism: Models and modeling perspectives on mathematics teaching, learning, and problem solving.* Mahwah, NJ: Lawrence Erlbaum Associates, Inc.

Lesh, R. (2001). Beyond constructivism: A new paradigm for identifying mathematical abilities that are most needed for success beyond school in a technology based age of information. In M. Mitchelmore (Ed.), *Technology in mathematics learning and teaching: Cognitive considerations: A special issue of the Mathematics Education Research Journal.* Melbourne, Australia: Australia Mathematics Education Research Group.

Lesh, R., & Carmona, G. (2003). Piagetian conceptual systems and models for mathematizing everyday experiences. In R. Lesh & H. M. Doerr (Eds.), *Beyond constructivism: Models and modeling perspectives on mathematics teaching, learning, and problem solving.* Mahwah, NJ: Lawrence Erlbaum Associates, Inc.

Lesh, R., & Doerr, H. (1998). Symbolizing, communicating, and mathematizing: Key components of models and modeling. In P. Cobb & E. Yackel (Eds.), *Symbolizing, communicating, and mathematizing.* Mahwah, NJ: Lawrence Erlbaum Associates, Inc.

Lesh, R., & Doerr, H. M. (Eds.). (2003a). *Beyond constructivism: Models and modeling perspectives on mathematics teaching, learning, and problem solving.* Mahwah, NJ: Lawrence Erlbaum Associates, Inc.

Lesh, R., & Doerr, H. M. (2003b). Foundations of a models and modeling perspective on mathematics teaching, learning, and problem solving. In R. Lesh & H. M. Doerr (Eds.), *Beyond constructivism: Models and modeling perspectives on mathematics teaching, learning, and problem solving.* Mahwah, NJ: Lawrence Erlbaum Associates, Inc.

Lesh, R., & Doerr, H. M. (2003c). In what ways does a models and modeling perspective move beyond constructivism? In R. Lesh & H. M. Doerr (Eds.), *Beyond constructivism: Models and modeling perspectives on mathematics teaching, learning, and problem solving.* Mahwah, NJ: Lawrence Erlbaum Associates, Inc.

Lesh, R., Doerr, H. M., Carmona, G., & Hjalmarson, M. (2002). Beyond constructivism. *Mathematical Thinking and Learning, 5*, 211–233.

Lesh, R., Hoover, M., Hole, B., Kelly, A., & Post, T. (2000). Principles for developing thought-revealing activities for students and teachers. In A. Kelly & R. Lesh (Eds.), *Handbook of research design in mathematics and science education* (pp. 591–646). Mahwah, NJ: Lawrence Erlbaum Associates, Inc.

Lesh, R., & Kaput, J. (1988). Interpreting modeling as local conceptual development. In J. DeLange & M. Doorman (Eds.), *Senior secondary mathematics education*. Utrecht, Netherlands: OW&OC.

Lesh, R., Post, T., & Behr, M. (1989). Proportional reasoning. In M. Behr & J. Hiebert (Eds.), *Number concepts and operations in the middle grades* (pp. 93–118). Reston, VA: National Council of Teachers of Mathematics.

Lesh, R., & Zawojewski, J. (1987). Problem solving. In T. Post (Ed.), *Teaching mathematics in Grades K-8: Research-based methods* (2003). Boston: Allyn & Bacon.

Lesh, R., Zawojewski, J., & Carmona, G. (2003). What mathematical abilities are needed for success beyond school in a technology-based age of information? In R. Lesh & H. M. Doerr (Eds.), *Beyond constructivism: Models and modeling perspectives on mathematics teaching, learning, and problem solving*. Mahwah, NJ: Lawrence Erlbaum Associates, Inc.

Lester, F. Jr., & Kehle, P. E. (in press). From problem solving to modeling: The evolution of thinking about research on complex mathematical activity. In R. Lesh & H. M. Doerr (Eds.), *Beyond constructivism: Models and modeling perspectives on mathematics teaching, learning, and problem solving* . Mahwah, NJ: Lawrence Erlbaum Associates, Inc.

Middleton, J. A., Lesh, R., & Heger, M. (2003). Interest, identity, and social functioning: Central features of modeling activity. In R. Lesh & H. M. Doerr (Eds.), *Beyond constructivism: Models and modeling perspectives on mathematics teaching, learning, and problem solving*. Mahwah, NJ: Lawrence Erlbaum Associates, Inc.

Piaget, J., Inhelder, B., & Szeminska, A. (1964). *The child's conception of geometry* (E. A. Lunzer, Trans.). New York: Harper and Row.

Post, T., Behr, M., & Lesh, R. (1982). Interpretations of rational number concepts. In L. Silvey & J. Smart (Eds.), *1982 Yearbook: Mathematics for the middle grades (5–9)* (pp. 59–72). Reston, Virginia: National Council of Teachers of Mathematics.

Van Hiele, P. M. (1986). *Structure and insight.* Orlando, FL: Academic Press.

Vygotsky, L. S. (1978). *Mind in society: The development of higher psychological processes.* Cambridge, MA: Harvard University Press.

Zawojewski, J., & Lesh, R. (2003). A models and modeling perspective on problem solving. In R. Lesh & H. M. Doerr (Eds.), *Beyond constructivism: Models and modeling perspectives on mathematics teaching, learning, and problem solving*. Mahwah, NJ: Lawrence Erlbaum Associates, Inc.

Zawojewski, J., Lesh, R., & English, L. D. (2003). A models and modeling perspective on the role of small group learning activities. In R. Lesh & H. M. Doerr (Eds.), *Beyond constructivism: Models and modeling perspectives on mathematics teaching, learning, and problem solving*. Mahwah, NJ: Lawrence Erlbaum Associates, Inc.

MATHEMATICAL THINKING AND LEARNING, 5(2&3), 191–210

Using a Modeling Approach to Analyze the Ways in Which Teachers Consider New Ways to Teach Mathematics

Roberta Y. Schorr
Department of Education and Academic Foundations
Rutgers University-Newark

Karen Koellner-Clark
College of Education
Georgia State University

This article describes how principles relating to modeling theory can be used to understand teacher development and the nature of teacher knowledge. We will document and analyze the actual practices of teachers who are attempting to modify, revise, and refine their approaches to the teaching and learning of mathematics, and then identify some of the conditions that have motivated the changes. Understanding teacher development in this way is of great importance because there are many efforts underway which attempt to reform the teaching and learning of mathematics. A primary objective of these efforts is to help teachers move toward the teaching practices established by the National Council of Teachers of Mathematics (NCTM, 1989, 2000). These practices entail a "radical change in the mathematics taught in schools, the nature of students' mathematical activity, and teachers' perspectives on mathematics teaching and learning" (Simon & Tzur, 1999, p. 252). But, reforming teachers' practice is not a simple matter. Stigler and Hiebert (1999), made the point that teaching is a cultural activity, and cultural activities "evolve over long periods of time … and rest on a relatively small and tacit set of core beliefs about the nature of the subject, about how students learn, and about the role that a teacher should play in the classroom" (p. 87). Stigler and Hiebert (1999) and other researchers underscored the difficulty in reforming deeply held and robust teaching practices.

Requests for reprints should be sent to Karen Koellner-Clark, College of Education, Georgia State University, Room 666, Atlanta, GA 30303. E-mail: kkoellner@gsu.edu

In the sections that follow, we examine some of the changes that are taking place in the teaching and learning of mathematics in the context of a large-scale study. We then describe a theoretical framework that can be used to explain the results of the study in terms of models and modeling. Next, we share a model for teacher development based on our theoretical framework, and provide further documentation and analysis to connect the changes that are taking place with the perspective on models and modeling. This documentation will include an in-depth analysis of a teacher in transition who is typical of the many teachers that we have worked with using this professional development approach and framework.

A CLOSER LOOK AT ACTUAL CLASSROOM PRACTICES

Spillane and Zeuli (1999) noted that many of the teachers involved in a study exploring patterns of practice in the context of national and state mathematics reform movements used many of the strategies often associated with the reform movement, such as concrete materials and small group instruction; however, they cite that the type of instruction that took place did not suggest to students that knowing and doing mathematics involved anything more than memorizing procedures and using them to compute right answers. In another large scale study, designed to document the teaching practices of fourth-grade teachers throughout the state of New Jersey (many of whom were involved in teacher development projects with colleges and universities), results suggest that teachers are indeed adopting some strategies that are often associated with reform (like using manipulatives, small group instruction, real-life problem activities) without substantially increasing the intellectual demands of their teaching (Firestone, Monfils, Camilli, Schorr, Hicks, & Mayrowetz, 2001; Schorr & Firestone, 2001).

In this particular study, 58 teachers were surveyed about their teaching practices, observed for at least two lessons, and then interviewed about each lesson. In all, 126 observations were done. The teachers came from districts throughout the state, representing the geographic, demographic, and socio-economic distribution of the state. The interviews and survey data reveal that the teachers feel that they were incorporating reform-oriented approaches into their teaching of mathematics. For instance, almost half of the interviewed teachers reported that they encouraged their students to explain their thinking; one third reported that they have students work with open-ended problems regularly; and almost one fifth reported that they are now de-emphasizing drill and computation.[1]

[1]During the 2000–2001 academic year, 55 teachers were interviewed. One sample of 23 fourth grade teachers was selected from among teachers who had completed the interview in 1999–2000. The other sample, collected for a related study, consisted of 32 fourth grade teachers in seven districts that received extensive professional development in elementary mathematics and science from New Jersey's State Systemic Initiative.

During interviews, teachers reflected on what they felt they learned from their respective professional development experiences. Several themes emerged. Many teachers mentioned that they learned to use manipulatives or hands-on materials. The teachers reported that they liked manipulatives because they felt that the materials motivated students and increased interest in mathematics. They also reported that as a result of their professional development experiences, they had learned to use a "constructivist" approach in their teaching. One urban teacher described what she meant by constructivist methods of teaching mathematics by stating the following:

> You know, I'm not a "here's the rule, plug it in, you know, practice it 100 hundred times" kind of math teacher. I love this, the exploration of ideas and questioning and proving and disproving and just that whole banter that goes on in the classroom when they're just starting to settle in and understand certain ideas.

Observation data, however, suggested that while teachers were reporting that they were using these techniques and strategies, the changes they have actually made are not nearly as substantive. As a case in point, we found that many of the teachers do indeed regularly use manipulatives. In fact, they were used in at least 60% of the observed math lessons. Yet, the fact that manipulatives were used does not mean that they were used effectively, or in a manner consistent with the intent of the standards. In fact, in most of the cases where they were used, they were used in a very algorithmic manner.

Few would argue that it is important for students to have "concrete" experiences in solving mathematical problems. However, when the manipulatives are used in an algorithmic way (i.e., when children use manipulatives as directed by the teacher, generally without understanding how and why the materials connect to the actual problem or other representations, including symbolic representations), children do not necessarily benefit from having used them (Lesh, Post, & Behr, 1987; Nobel, Nemirovsky, Wright, & Tierney, 2001). Nobel, Nemirovsky, Wright, and Tierney (2001) appropriately noted that the concepts and ideas do not reside in the physical materials, nor in the prescribed classroom activities, but in what students actually do and experience. When a teacher tells the students exactly how to use manipulatives (or other types of materials or activities), and the students must promptly follow along as directed by the teacher, the opportunity for exploration is often lost, as the students are merely following a procedure—which happens to be manipulating materials. Unfortunately, this was the case in most of the lessons that were observed where manipulatives were used. As an example of this, consider one lesson where the teacher attempted to have students solve a problem using concrete materials—chips—while working in a small group setting. The teacher placed students in small teams (groups) of four to work together to solve the following problem: "There are 84 fourth graders, and because they've done so well, [I have] decided to take 2/3 of them out to dinner with her...so I'm trying to find out what is 2/3 of 84." She distributed chips to the students, and told them to work on the problem using the chips—and not using paper and pencil. Many of the students began to separate the chips into 4 different groups, which was not the strategy that the teacher had expected. After a very short period of time she decided to tell the students exactly how to arrange the chips. She instructed them that they needed "three groups, because two thirds means two out of three groups. So if you have a pile of 84, then you need to make three groups, and you keep passing them out into groups until you've used up the 84. Remember the denominator tells you the number of equal groups you need. Just like if you play cards, and each person gets the same number of cards, right." After demonstrating this procedure, she went around to each group to be sure that they had done it exactly as she had prescribed. When the students were finished, she continued as follows:

T: OK, we finished step one. If you want to know what two thirds of 84 is, you have to divide 84 by 3. So how many did you end up with in each group?

Girl: 28

T: So what do we do now? How do you know what type of equal groups to put them in? What tells you?

Boy: By looking at two thirds?

T: What tells you? The denominator tells you what number of equal groups to divide by. The divisor or the bottom number of the fraction tells you how many equal groups to make. So does that mean that 28 students can go? Can I get a consensus? (About one half of the students raise their hands). 28 people cannot go. So what do I need to do now? Two thirds of my 84 students can go. So how many students can go? You're not multiplying; you're using your manipulatives. So what am I doing now? Everyone should have the same amount.

Notice that the teacher directed the students to make three groups of chips. This was done with little further discussion of, for example, why three groups were needed in this particular problem, or why the three in the two thirds is used to determine the number of groups of chips that the children should have, nor did she elicit why some groups had originally made groups of four. While the teacher did say that the denominator tells what number of equal groups to divide by, there was no discussion as to why or if that is always the case, nor whether there were other options, or how that maps into the concrete representation. In fact, there was evidence of confusion throughout the entire lesson. For example, many of the teams of students initially were dividing the chips into four piles rather than three (recall that the students worked in teams of four). Many students appeared to not have a clear (if any) understanding of why the teacher felt that three groups were needed—or indeed if three groups were needed.

Despite the fact that the students had placed the chips into three groups (as instructed), they were still not able to find the solution to the problem. As a result, this teacher allowed a student to demonstrate a written algorithm for division on the board. The student divided 84 by 3 thereby obtaining a quotient of 28. The teacher then instructed the students to put all of the chips back into one pile again, and redistribute them so that now they would have only three chips in each group (28 groups of 3). In an attempt to justify this decision, she said the following:

> The whole reason I let you put them into groups of three [referring to the first part of the lesson where the chips were distributed into 3 piles of 28] is that I wanted you to see that when you are dividing, you need to look at the dividend, that is the number I have in all, and the divisor tells me how many in each group...What if I have 10 party bags and 5 people coming to my birthday party? How many bags would each person get? Each person would get 2 bags. The divisor tells you how many in each group, and the quotient tells you how many groups. I told you 2 from each group can go. So now tell me, how many can go. Take 2 from each group on a blank area of you desk. Now tell me how many people can I take out to lunch.

Most of the students were still not able to come up with the answer, and so the teacher again directed them as to how to distribute the chips, and solve the problem.

Throughout the lesson, the teacher was instructing the students to distribute the chips in a way that appeared to make sense to her. Her directions indicated that she was considering different models for the solution in the latter part of the lesson than she had in the beginning. As one might expect, most of the students in that classroom did not appear to understand how the different representations connected to each other, the algorithm, or to the problem activity. Furthermore, the teacher did not appear to ever notice that. For example, in her post lesson interview, she acknowledged that there was some confusion, but felt that the lesson went well: "I think the manipulatives and the hands-on experience worked well, and I think the cooperative groups with them working together and learning from each other worked well." The interviewer specifically asked her about the different ways in which she instructed the students to use the materials. She indicated that she had a specific goal for the lesson, and to accomplish it, she needed to have the students sort the chips in a particular manner.

The purpose of sharing this excerpt is to document that while many teachers had students physically touch concrete manipulatives, there often was little or no opportunity for the students to actually develop their own solutions to the problem, connect the materials to the problem activity, or consider how the concrete representations might be connected to symbolic or other representations. Data from this study also suggests that while teachers may use the language of reform, their actual classroom implementations did not reflect more than surface level types of activities. In fact, the tasks that were actually implemented almost always involved either straight memorization and/or performing procedures. Only 2.5% of all sessions involved situations where students were required to do higher-level thinking. While teachers were reporting that they were encouraging students to share ideas about solution strategies, this did not happen very often. Teachers would say "I encourage them to understand that there may be more than one solution to a particular problem, and I would encourage them to use a variety of methods to solve a particular problem" or "I want students to explain their answers and understand why and how it works", but in most cases, students did not explain their thinking. In fact, students were only observed to explain the thinking behind the approach they used in seven lessons (6%).

Overwhelmingly, the data suggest that teachers are still primarily interested in having students get the correct answers on relatively routine and primarily procedural problems, and with very little in-depth elaboration on their solution strategies. Even when students did share ideas or solutions, their responses were usually limited to a short neutral comment like "okay" or "fine," and no attempt was made to get the student to expand on the answer or use the answer to start a discussion. Students were rarely challenged to consider whether the answer given was reasonable (in 79% of all cases students were not challenged). Furthermore, it was found that when a student asked an interesting question or raised an interesting issue, teachers often ignored it, or responded to it in a superficial way. Indeed, the type of conversation that took place did not lead to the conceptual understanding of big ideas. In fact, procedural solution strategies appeared to remain the dominant feature. Teachers noted that many they had to use procedural approaches, citing that this was necessary because the students were having trouble understanding what to do, and they did not feel that the students could understand it any other way.

The above data suggest that there is still an overwhelming emphasis on procedures in these mathematics classrooms, and meaningful discourse is clearly not the norm. Furthermore, these data indicate that teachers do not appear to be focusing on the conceptual development of knowledge. These results are not unique. Koellner-Clark, Bote, and Middleton (1998) found that a first grade teacher who claimed to be focused on reform oriented assessment strategies aligned with Cognitively Guided Instruction continually explained that she "listened to student thinking to assess students." Indeed, she did record solution strategies such as skip counting, counting all, etc., as she watched her students' work. However, her ability to correctly identify the counting strategies did not imply that she was able to use them to inform her teaching, or to make any other types of instructional decisions. In fact, when considering how to assess the children's work, she would grade them as being either right or wrong.

Simon and Tzur (1999) similarly noted that "Although teachers have been able to appropriate from the reform movement particular teaching strategies (e.g., using small groups, manipulatives, and calculators), the movement has not provided them with clear direction for how to help students develop new mathematical ideas" (p. 258). While it may appear that adopting surface level strategies or techniques may not be particularly effective, Stigler and Hiebert add an additional word of caution by stating that "trying to improve teaching by changing individual features usually makes little difference, positive or negative. But it can backfire and leave things worse than before"(p. 99).

In the sections that follow, we describe, using a modeling perspective, why the deeper aspects of reform are not being incorporated into actual teaching practices more regularly, and we also provide examples of how teachers can build new models for teaching and learning which do indeed move beyond the more surface level characteristics cited above. To do this, we will explain our underlying philosophical perspectives regarding conceptualizations about the teaching and learning of mathematics in terms of models and modeling.

USING A MODELING PERSPECTIVE TO UNDERSTAND TEACHER DEVELOPMENT

In this section, we describe how a modeling perspective can be used to understand teacher development. We consider a model to be a way to describe, explain, construct or manipulate an experience, or a complex series of experiences. Models are organized around a situation or an experience. A person interprets a situation by mapping it into his or her own internal model, which helps him or her make sense of the situation. Once the situation has been mapped into the internal model, transformations, modifications, extensions, or revisions within the model can occur, which in turn provide the means by which the person can make predictions, descriptions, or explanations for use in the problem situation. Models help us organize relevant information and consider meaningful patterns that can be used to interpret or reinterpret hypotheses about given situations or events, generate explanations of how information is related, and make decisions about how and when to use selected cues and information. Models, according to our hypothesis, develop in stages where early conceptualizations may be fuzzy, or even distorted versions of experience, and several alternative models may be available to interpret a given problem situation. We would speculate that models that have been in use for long periods of time, and appear to be effective for a given purpose are, generally speaking, quite robust.

According to this perspective, all teachers have "models" for teaching and learning mathematics. These models indeed govern the ways in which teachers actually teach. While teachers may express new beliefs about children, teaching or learning, unless fundamental changes in their models occur, their practice will remain relatively unchanged—the study above provides documentation for that. The models that teachers possess have been established as a result of their own experiences both as a student and as a teacher. These models are built around the teacher's own:

- learning experiences involving mathematics (i.e. how the teacher experienced mathematics as a student),
- understanding of the mathematical concepts involved,
- understanding of how children learn the concepts involved,
- understanding of the pedagogical practices that will enable children to "learn" the concepts involved and,
- experiences with materials, curriculum, and other resources, including technological resources.

When teachers adopt specific changes or strategies (like using manipulatives) into their classroom practice, they often do so within the framework of their older, (more traditional) models. That is, they incorporate a new technique or strategy into their already existing model for teaching and learning. The new technique or strategy is added onto their model, without fundamentally changing their worldview of what mathematics instruction is or should be. Data from the study cited previously (Schorr & Firestone, 2001) documented instances of this, and confirm that teachers are readily willing to adopt a new strategy, tactic, or procedure, but, generally speaking, do so without changing their overall perspectives about the teaching and learning process.

To illustrate this notion, consider the stable and robust model as a sphere (see Figure 1). The implementation aspects of reform that we have described above are illustrated by the smaller circles attached to the outer edge of the sphere. The modifications resting at the edge of the circle have not penetrated the sphere, or fundamentally altered the model that the teacher holds for teaching and learning. According to our hypothesis, this is because the teacher has incorporated only certain "surface" level aspects of the reform into his or her teaching practices, rather than a new world view of the teaching and learning process.

To revise, refine, or reject previously held models, a teacher would need to be engaged in experiences that would, metaphorically speaking, provide him or her with an opportunity to "break through" the surface of the previously held model(s). These experiences must cause the teacher to encounter cognitive conflict or dissonance—which in turn challenges the previously held notions. Not surprisingly, the same can be said of any learner—whether that learner is a child learning about fractions, or a teacher considering new ways to teach the content. When a learner experi-

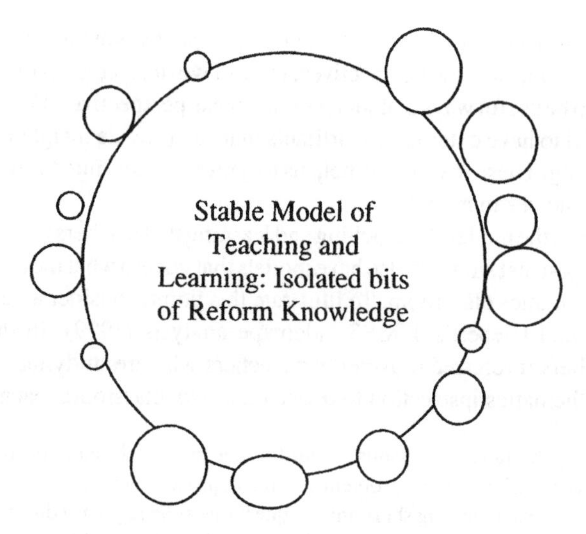

FIGURE 1 A Robust Model of Mathematics Teaching and Learning with Minor Attachments of Isolated Pieces of Reform Oriented Knowledge.

ences a difference between the perceived outcome of an event or problem, and the predicted outcome, he or she is pushed to reconsider the foundation on which the prediction was made, and make revisions accordingly . Overall, deep changes occur when the teacher has considered, at a deep level, many things, including the mathematical ideas involved, the ways in which students build these ideas, and the impact of his or her pedagogical decisions on student learning. When the teacher has had such opportunities, he or she is then able to use them as a basis for revising older, more robust models.

When a teacher experiences perturbations, and then has the opportunity to reflect on them, modifications and revisions within the model can begin to take place. It is these revisions within models that can begin to, again, metaphorically speaking, change the shape and structure of the model. We would suggest that such revisions occur over time, and in stages. Early conceptualizations or stages are often fuzzy versions of later stages. The fundamental changes that occur in the model lead to new worldviews regarding the entire teaching and learning process.[2]

[2]Knowledge development according to this perspective can be understood in terms of changes in a learner's model or conceptual system. A learner's current state of understanding or his conceptions are impressed into his internal model based on his experiential reality, and his interpretations of it. We do not claim that these ideas are new. Indeed they are consistent with the Piagetian notion of constructivism (Piaget, 1970, 1980) whereby human beings incorporate new understandings into their existing sets of understandings as a result of their experiences with the world. We would also maintain that these ideas are consistent with notions relating to social enculturation (Cobb & Yackel, 1996) in

We cannot draw conclusions about teachers' perspectives or their conceptions because we cannot know these perspectives; instead we draw conclusions about what we consider to be useful ways of understanding these perspectives. For this reason, it is most helpful to have externalized artifacts that can provide insight into teachers' ways of thinking. These in turn, can help us to gain a window into knowledge development and changes in models.

In addition to the models for teaching and learning that teachers bring to the classroom, we suggest that students also have models that govern what they expect to happen in a mathematics classroom. To illustrate this point, consider an excerpt taken from Stigler and Hiebert's TIMSS videotape analysis (1999). In this example, Stigler and Hiebert referred to American teachers who are studying videotapes of Japanese mathematics instruction to revise their own classroom teaching.

> After viewing the Japanese lessons, a fourth-grade teacher decided to shift from his traditional approach to a more problem-solving approach such as we had seen on the videotapes. Instead of asking short-answer questions as he regularly did, he began his next lesson by presenting a problem and asking students to spend 10 min working on a solution. Although the teacher changed his behavior to correspond with the actions of the teacher in the videotape, the students, not having seen the video or reflected on their own participation, failed to respond as the students on the tape did. They played their traditional roles. They waited to be shown how to solve the problem. The lesson did not succeed.

From the above excerpt, it would appear that this teacher attempted to synthesize a relatively small aspect of an entirely different approach to teaching and learning mathematics into his preexistent, and his student's preexistent, models for teaching and learning mathematics. As one might predict both responded based on the older, more traditional models.

In the previous section, we have documented instances where teachers have not fundamentally altered their models for teaching and learning. In the next sections, we will document instances where, using a multi-tiered teacher development design, teachers are able to adopt new models for teaching and learning mathematics.

MULTI-TIER PROGRAM DESIGN

Our research has documented the benefits of a multi-tier program design in helping participating teachers to develop the knowledge and ability to teach mathemat-

---

that learners are participants in social groups or communities, in our case workshop groups, wherein each workshop group is characterized by the norms, values, practices and shared knowledge they socially construct. Knowledge development takes place in a coordinated system of social and cognitive processes that occur simultaneously. Similarly, knowledge development, from a modeling perspective, is a simultaneous function of social and cognitive processes coordinated with multiple opportunities in which teachers are able to construct, explain, design, or organize a series of experiences in a thoughtful way.

ics differently. Multi-tier program designs focus on the interacting development of students, teachers and researchers (see Figure 2). In this design, students (as well as teachers) are engaged in problem-solving situations that repeatedly challenge them to reveal, test, refine and revise and document important aspects of mathematical constructs. As this occurs, teachers are involved in problem-solving experiences that focus their attention on their students' modeling behavior. Teacher level problem-solving experiences take several forms. For example, one type of problem activity might require teachers to construct a tool for classroom decision making like a "ways of thinking sheet" to document student strategies, conceptions and misconceptions as the students are working (as described by Schorr & Lesh, 2001). Another activity for teachers could be to design observation sheets or concept maps, which are focused on their students' mathematical thinking (as described by Koellner-Clark & Lesh, 2001). In all cases, teachers are encouraged to produce conceptual tools, which are also sharable or reusable. Teachers can use these tools to make predictions about actual classroom outcomes, and then refine and revise accordingly. Researchers and/or teacher educators are simultaneously focused on the nature of teachers' and students' developing knowledge and abilities—all of which affect each other.

Overall, the problem tasks that are used—whether at the teacher level, or at the student level, are intended to provide a means by which teachers can examine and reflect on their students' modeling behavior, the mathematical ideas involved in their curriculum, the ways in which students build these ideas, and the impact of their pedagogical decisions on student learning. All experiences are also intended to encourage teachers to reflect on their mathematical content and pedagogy and discuss their ideas within a community of peers. (Note the distinct difference between this approach and the more typical expert/novice approaches in which ex-

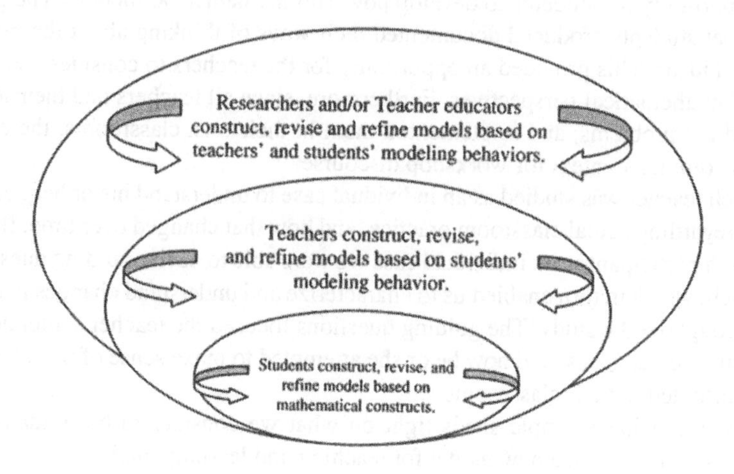

FIGURE 2 Multi Tier Program Design.

perts come in with a preconceived notion of what a "good" teacher should be and then "tell" teachers how to get to that state.)

CHANGING ACTUAL PRACTICES

We now describe a case in which we document the development of new models in teachers. In this particular case, twelve middle school teachers had been participating in a 16-week multi-tiered program design project. In this project, the participating teachers agreed to implement problem-solving activities within their classrooms, and come together to share ideas. Each week they, along with a researcher, worked in small groups to discuss:

- the mathematical content present in the problem;
- student strategies and misconceptions they identified during implementation;
- the mathematical discourse they documented from their classroom;
- their reflections regarding their pedagogical decisions; and,
- video clips taken from their classrooms.

Researchers also observed the teachers as they implemented the problem activities in the context of their own classrooms. It was these workshop discussions and classroom observations that serve as the basis for our identification of changes in the teaching practices of the teachers.

The problem-solving activities that were used with the teachers and students were considered to be, "thought revealing," in that they were intended to provide an opportunity for students to develop powerful mathematical models. The products that students produced documented their ways of thinking about the mathematical ideas. This provided an opportunity for the teachers to consider their students' mathematical perspectives. Furthermore, since all teachers had themselves solved the problems, and implemented them in their own classrooms, there was also a common context for workshop discourse.

Each teacher was studied as an individual case to understand his or her perceptions regarding actual classroom practice, and how that changed over time. By using each participant as an individual case we were able to devise guiding questions for each, which in turn enabled us to characterize and understand changes in models throughout the study. The guiding questions focused the teacher's attention to reform-oriented ideas, and how he or she attempted to make sense of the situation as it unfolded in their classrooms.

The following example sheds light on what we consider to be fundamental changes that result in a new model for teaching and learning mathematics.

A Case Study of Roger

From the outset of our study our guiding questions regarding Roger were:

- What are Roger's current approaches to the teaching and learning of mathematics?
- What reform-oriented practices is Roger attending to and what are the ways in which he implements them?
- How does Roger promote changes in students' mathematical knowledge?
- What should the researcher consider to provoke Roger to consider new instructional practices (if appropriate)?

We use these themes in our analysis of Roger. They also provide the foundation for considering changes in Roger's model(s) for teaching and learning mathematics.

Roger was in a workshop group of five teachers. From the outset, he shared a desire to use his students' understanding as a basis for his instructional decisions. Like the teachers in the previously cited study, Roger continually expressed the opinion that it was important to "listen to student thinking" and "have students share their solutions." On numerous visits to his classroom, researchers noted that he did indeed ask his students to explain their thinking. However, Roger, like many of the teachers in the study cited above (Schorr & Firestone, 2001) did not use student thinking as a means for encouraging substantive conversation, or as the basis for instructional decisions. The explanations that the students gave were procedural in nature and Roger did not use student solutions to spark discussion or argumentative debate, discuss the reasonableness of answers, or help students make conjectures relating to other mathematical ideas. This is despite the fact that all of these ideas had been discussed in workshop groups and identified as important for teachers interested in listening to student thinking and having students explain their solutions. According to our models and modeling perspective, this is somewhat predictable in that teachers, like their students, cannot be expected to build new models for interpreting their experiences by simply listening to others talk about the ideas. Knowing when and why to use a given idea is quite different than knowing how to use it.

In an effort to better understand what Roger meant by "listening to student thinking" or "having his students share their solutions," we asked Roger to discuss these ideas during interviewed sessions. He told us that "the NCTM standards promoted this idea of 'communication' and to him if students could share how they solved a problem then other students might be able to learn from that [student's] explanation." We followed up with a probe asking him to describe the types of explanations that he might be looking for. "I am looking for different solutions, if students solve problems differently these solutions should be shared." It appeared from his words and actions in the classroom that Roger felt that having students explain their mathematical thinking was important because it enabled students to listen to, and learn from each other. However, Roger, like all learners, needed to have an opportunity to consider these ideas, "test" them and then refine and revise them accordingly.

We would suggest that Roger knew about, and attempted to adopt what he identified as a reform-oriented perspective regarding discourse into his classroom instruction. However, our data indicate that this was done within an older, more traditional model (or way of thinking) about teaching and learning mathematics. In an effort to help him revise his models, we sought to learn more about what Roger attended to as he listened to his students and how he used the information obtained from his student's responses. During workshops, he continually stressed that he used student thinking as a marker for progress. He also maintained that students learned from each other and therefore it was important to have students share their ideas. His peers in the workshop group were also interested in learning more about why student thinking was so important to him. When asked by his peers, he responded as follows "I don't write student solutions or thinking down but over the years I almost can predict the different ways my students will solve particular problems that I have used more than once." (Workshop Transcript). To Roger, the usefulness of student thinking for the purpose of assessment was summed up as follows:

> I know where the students are in their thinking by how they respond. It is like an informal assessment that I use to decide what to teach the next day. I guess it is like a barometer for me to make sure they got it and I don't need to go over it again, you know, in a different way. It is a marker of progress. (Workshop Transcript)

In-class observations confirmed that Roger did not use this information systematically to make informed instructional decisions, and he did not use it to delve deeper into his students' conceptual understanding (as discussed in the workshop group) rather, he used the information "to make sure they got it."

In terms of a models and modeling perspective, we speculate that at this point in time Roger had not yet altered his model for teaching and learning. Rather, Roger's idea of asking questions to better understand his students' thinking was appended to his more traditional model of instruction. Evidence of this can be found in that Roger asked his students how they solved particular problems but only attended to the more procedurally oriented aspects of their strategies to decide whether they "got it."

In an effort to better understand Roger's ways of thinking, he was asked to share a videotape with the workshop group that exemplified his approach to using student thinking. He chose a clip that he felt illustrated his ability to mark progress of particular students. The following excerpt will document the clip that he was referring to. In this particular case, the students were working on a problem involving ratios and proportions.

Roger: "Sonya, could you explain to the class how you decided that the package A was the best deal? Tell us a little about the mathematics you used to solve the problem."
Tanya: "I took 1/3 times 68 and 1/4 times 90 and then 2/3 times 48,"
Roger: "Can you explain in more detail?"
Tanya: "Yes, I took 68 divided by three which was 22 2/3. Then I took times 90 again by dividing by 90 by 4, which was, 22, and then I took 2/3 times 48 by multiplying .66 times 48, which was 31.68. So I knew that package A was the best deal."

Roger noted that he chose this excerpt to show

> how my students are really used to explaining their answer. They do it naturally almost. I am impressed with how Sonya can explain the procedures to the class in such an organized fashion. She has a good conceptual understanding of this problem… I can always count on her to share her solutions because she is good at explaining them too. (personal communication, interview)

At this time, the group discussed Sonya's mathematical understanding as evidenced on the videotape. Some colleagues asked Roger if he felt that "Sonya understood the estimated values and the proportionality of the situation". This discussion went on for about 20 min with Roger, confidently explaining his interpretation of Sonya's thinking. As the questions kept coming Roger became less confident and clear about how Sonya interpreted the problem. While he felt that she had procedurally and succinctly described her solution, he was no longer certain that he knew enough about her conceptual understanding of the problem.

Roger now noted that he apparently was not asking the right questions. He shared that he really did want to understand how the students interpreted the problem mathematically, but that although he was asking students to explain their answers he was still emphasizing the procedures in the answer and whether the solution correct. "I guess I need to think about what parts of proportionality I am focusing on because really Sonya was just showing the procedure to get the right proportion" (Workshop Transcript). This insight helped Roger and the other participants begin to analyze their teaching practices in a deeper way. For Roger, this represented the first time that he began to fundamentally question his approach to instruction. We would suggest that at this time, he began to revise his models for teaching and learning.

From this point on, Roger made a decided effort to shift the focus of his classroom discussions from procedures to larger conceptually based ideas. He struggled with this notion as it touched on all facets of his teaching. He began to question his objectives, his lesson plans, the activities and problems he chose, and the things his students said during class discussions. He found that they were all interrelated.

> In order for students to discuss solutions in a way that I can help them talk about big ideas I first have to give them problems so that this can happen. I mean sometimes I give them problems and they really don't lend themselves to this kind of discussion. So when I started to reorganize my lessons I realized that I didn't even know what kinds of things [activities] would help them understand proportions. It was different stuff than I had in my file cabinet because that stuff was just "cross multiply and divide stuff." So I keep trying to "hear" the discussion that might take place when I give my students different problems to solve because I figure when I organize my curriculum-unit the way I want, then I will be able to "hear" a discussion that I can facilitate to make sense of big idea. (personal communication, interview)

Notice that in the above excerpt, Roger attempted to imagine how his students might solve the problems before he actually used them in class. These types of reflections caused him to (re)consider many things including the mathematical ideas involved, how they are developed in the curriculum, how they are understood by students, and what role he might play in the process. As he grappled with these ideas, he shared less with his peers during workshop settings. He did, however, continue to reflect on these ideas in his written journals. He emailed the researchers more frequently to ask questions or receive feedback about his ideas, experiences and instructional decisions. For instance, he often asked questions about activities that he was preparing for the following day based on his students' work, including their responses, strategies and mathematical thinking that he had documented from that day. He shared his own struggle with trying to understand the big conceptual ideas. For example, as he questioned his instructional decisions he made it clear that he was questioning his own understanding of the skills, activities, and questions that would help his students learn the big idea intended.

Not surprisingly, he also shared that the more he changed his instructional practices, the more his students changed. They began to debate with each other about the reasonableness of an answer. Roger pointed out that watching students try to make sense of each others solution provided him with a great deal of insight as well. Roger used more open-ended problems that provided an opportunity for students to explain their thinking. This was important for Roger in that he could begin to understand how his students thought about the mathematical ideas. This, in turn, led him to think about how he could use student solutions to further the mathematic discourse in his classroom.

Later in the year, Roger shared another video clip with the workshop group. His peers noticed the changes that had taken place, and appropriately commended him on the progress he had made with facilitating discussion and building on students' existing understandings of mathematical concepts. Roger shared his interpretations of his students' understanding of the mathematical ideas involved in the problem they were solving. The workshop group carefully analyzed the mathematical solutions of two students. They questioned Roger to better understand the pedagogical decisions he had made. For example, they asked him about his choice of questions, his decisions about which students to ask to share their solutions, and the sequence of problems that he used before and after the one used on the video. All of Rogers's responses were carefully considered. He noted that he had to constantly refine and revise his approach to promoting thoughtful discourse. For example, he stated that prior to this occasion, he had not necessarily given much forethought to which students he might choose to share solutions. He stated

> You know I guess I am not waiting for particular strategies to be used to capitalize on their thinking to benefit the class discussion—I was just letting the students share their strategies with each other. I need to think about the implications of using student thinking or their solution strategies to aid in the debate of class discussion. (Workshop Transcript)

These remarks indicate that he intended to carefully consider certain student strategies as he attempted to elicit a discussion in which students could defend and justify their work.

Roger continued to reflect on his teaching, and revise and refine his models accordingly. One year after the last workshop session, Roger still maintained that he was attending to his students' thinking to make informed instructional decisions.

> I always try and think about how the students are interpreting the problem whether it is a fraction problem or a problem solving activity. I feel like I have insight into what my kids know and some of the misconceptions or holes or connections that they don't see so I am able to help my kids learn mathematics in a better way. I must say that not collaborating in the workshop has probably slowed me down. I think my group really helped me understand my content and my teaching better but I don't think I have lost sight of what I learned. (personal communication, interview)

As he reflected on the entire workshop experience, he recalled the instance where he had shared a video clip of his classroom:

> I will never forget one of the first times I shared a video clip of my students "sharing their thinking" and all they were really doing was coming to the board and writing the steps to a problem down. They really weren't learning much from each other or about mathematics but I thought I was being an effective mathematics teacher. (personal communication, interview)

When asked what was an effective mathematics teacher Roger explained,

> ...I guess someone who knows how kids think about mathematics but realizes that they do think about it differently. You cannot always know how they are thinking so you should always remember that and teach in light of that. I try and listen to my kids individually and when they work in groups. I try and figure out how to proceed for the day's lesson as well as the next day's lesson based on what I heard from them. I must be honest though, I do do this but I do not do this everyday because too many things are happening everyday. I would say that I believe this is what I should do and I do it most of the time but like recently the kids were getting ready to test [standardized testing] and although I tried to review with them from a conceptual way I don't think reviewing can really be done without traditional procedures and rules. I did remind them of previous things we did to help them understand particular concepts but anyway, testing isn't really aligned with how I teach math. (personal communication, interview)

Roger was observed twice that same week. His approach had indeed become much more thoughtful. For example, in one case, the students were engaged in a problem activity where they were discussing appropriate statistical methods to analyze data. Many of his students had chosen the mean to describe the center of a set of data. Roger effectively questioned the students about their decision to use the mean. This lead to a thoughtful discussion where students began to use other ideas as a measure of central tendency.

During the interview that took place at the conclusion of the lesson, he discussed the activities and experiences that he would be implementing to help his students build and further their understanding about the measures of central tendency. His remarks about the activities that he had chosen were thoughtful and showed how his mathematical and pedagogical knowledge had continued to evolve.

CONCLUSIONS

In this article, we have provided examples of teachers who have adopted new strategies, but not substantively changed their models for teaching and learning. We have also documented an instance when the changes are more substantive. In each case, we have analyzed the practices in terms of a models and modeling perspective. This perspective is useful in considering the conditions that are necessary for generating fundamental changes in practice.

Roger is an example of a teacher who is making the transition from a more traditional model of teaching and learning toward a more reform-oriented one. He is one of many teachers involved in reform efforts using this design. The multi-tiered program in which he participated provided him with the opportunity to consider his approach to teaching, make predictions about what was happening, test those predictions, and then reflect the outcomes in a collegial setting. It also allowed him to consider how his students think about particular mathematical ideas. Roger and the other teachers in the workshop group had a common context in which they could discuss content and pedagogy. In the process, they were able to reflect on their ideas about mathematics teaching and learning as they considered the strengths and weaknesses in their ways of thinking.

This type of experience, we believe, provided the impetus for teachers to develop new world-views about their teaching practices. As one would expect, all participants developed in ways that were unique, but we would maintain, progressively better. One commonality that did emerge from our analysis of all participants was that all teachers began to reflect more deeply about their students' thinking. The teachers asked more questions and closely listened student's responses.

It is through experiences like those described above that teachers can begin to "break" through their stable and relatively robust models for teaching and learning.

REFERENCES

Cobb, P., & Yackel, E. (1996). Constructivist, emergent and sociocultural perspectives in the context of development research. *Educational Psychology, 31*(3), 175–191.

Firestone, W. A., Monfils, L., Camilli, G., Schorr, R. Y., Hicks, J., & Mayrowetz, D. (2001). The ambiguity of test preparation: A multimethod analysis in one state. Manuscript submitted for publication.

Koellner-Clark, K., & Lesh, R. (2003). A modeling approach to describe teacher knowledge. In R. Lesh & H. M. Doerr (Eds.), *Beyond constructivism: Models and modeling perspectives on mathematics teaching, learning, and problem solving*. Mahwah, NJ: Lawrence Erlbaum Associates, Inc.

Koellner-Clark, K. A., Bote, L. A., & Middleton, J. A. (1998, April). *Cycles of transformation in assessment practices in a cognitively guided instruction classroom.* Paper presented at the Annual Meeting of the American Educational Research Association, San Diego, CA.

Lesh, R., Post, T., & Behr, M. (1987). Dienes revisited: Multiple embodiments in computer environments. In I. Wirszup & R. Streit (Eds.), *Developments in school mathematics education around the world* (pp. 647–680). Reston, VA: National Council of Teachers of Mathematics.

National Council of Teachers of Mathematics. (1989). *Principles and standards for school mathematics.* Reston, VA: Author.

National Council of Teachers of Mathematics. (2000). *Principles and standards for school mathematics.* Reston, VA: Author.

Nobel, T., Nemirovsky, R., Wright, T., & Tierney, C. (2001). Experiencing change: The mathematics of change in multiple environments. *Journal of Research in Mathematics Education, 32*(1), 85–108.

Piaget, J. (1970). *Genetic epistemology.* New York: Columbia University Press.

Piaget, J. (1980). *Adaptation and intelligence: Organic selection and phenocopy* (S. Eames, Trans.). Chicago: University of Chicago Press.

Schorr, R. Y. (2000). Impact at the student level. *Journal of Mathematical Behavior, 19,* 209–231.

Schorr, R. Y., & Firestone, W. A. (2001). Changing mathematics teaching in response to a state testing program: A fine-grained analysis. Paper presented at the Annual Meeting of the American Educational Research Association, Seattle, WA.

Schorr, R. Y., & Lesh, R. (2003). A models and modeling perspective on classroom-based teacher development. In R. Lesh & H. M. Doerr (Eds.), *Beyond constructivism: Models and modeling perspec-*

tives on mathematics teaching, learning, and problem solving. Mahwah, NJ: Lawrence Erlbaum Associates, Inc.

Schorr, R. Y., & Lesh, R. (in press). A models and modeling perspective on classroom-based teacher development. In R. Lesh & H. Doerr (Eds.), *Beyond constructivism: A models and modeling perspective on teaching, learning, and problem solving in mathematics education*. Hillsdale, NJ: Lawrence Erlbaum Associates, Inc.

Simon, M. A., & Tzur, R. (1999). Exploring the teacher's perspective from the researchers' perspectives: Generating accounts of mathematics teachers' practice. *Journal of Research in Mathematics Education, 30*, 252–264.

Spillane, J. P., & Zeuli, J. S. (1999). Reform and teaching: Exploring patterns of practice in the context of national and state mathematics reforms. *Educational Evaluation and Policy Analysis, 21*(1), 1–27.

Stigler, J. W., & Hiebert, J. (1999). *The teaching gap: Best ideas from the world's teachers for improving education in the classroom*. New York: The Free Press.

MATHEMATICAL THINKING AND LEARNING, 5(2&3), 211–233

Beyond Constructivism

Richard Lesh
School of Education
Purdue University

Helen M. Doerr
Department of Mathematics
Syracuse University

Guadalupe Carmona and Margret Hjalmarson
Department of Education
Purdue University

In a recent book titled *Beyond Constructivism: A Models & Modeling Perspective on Mathematics Problem Solving, Learning & Teaching* (Lesh & Doerr, 2003a), the concluding chapter describes a number of specific ways that a models and modeling perspective moves significantly beyond the implications that can be drawn from constructivist theories in the context of issues that are priorities to address for teachers, curriculum developers, or program designers. In that chapter (Lesh & Doerr, 2003b), the following topics were treated as cross-cutting themes: (a) the nature of reality, (b) the nature of mathematical knowledge, (c) the nature of the development of children's knowledge, (d) the mechanisms that drive that development, (e) the relationship of context and generalizability, (f) problem solving, and (g) teachers' knowledge and the kinds of teaching and learning situations that contribute to the development of children's knowledge. In this article, we organize our comments directly around the preceding topics and describe how a models and modeling perspective provides alternative ways of thinking about mathematics teaching and learning that enable teachers, researchers and others to produce useful and sharable conceptual tools that have powerful implications in the context of decision-making issues that are of priority to practitioners.

Requests for reprints should be sent to Richard Lesh, School of Education, LAEB 1440 Room 6130, Purdue University, West Lafayette, IN 47906. E-mail: rlesh@purdue.edu

Nearly all constructivists agree that knowledge is actively constructed by the student and not simply passively received from the teacher. Indeed, hardly any modern theories of learning would argue students passively receive knowledge from teachers. Similarly, most constructivists agree that knowing is not the discovery of some objective and preexisting world. However, in many important decision-making areas, the claim that "knowledge is actively constructed by the learner" is of limited use to teachers and researchers, especially if details are lacking about how the constructs are to be developed. For example, a number of recent publications have shown that teachers and researchers claiming to be constructivists can hold exceedingly different views about practical issues such as: the kinds of examples and representations to use, the role of concrete materials in instruction, the quantities and types of guidance and feedback to give students, and strategies for the use of computational tools and environments (Goldin, 2001; Kelly & Lesh, 2000; Steffe & Wood, 1990). At issue is not whether the foundational claims of constructivism are "true" or "false" in some absolute sense, but rather, whether the basic principles are meaningful and useful to teachers, curriculum developers, researchers and other educational decision makers when compared with other competing theoretical perspectives.

THE NATURE OF REALITY

We begin by focusing on the differences in the starting propositions that embody important distinctions between a modeling perspective and a constructivist perspective on the relationship between the individual's experiences and the nature of reality. Radical constructivism is primarily a theory designed to address issues that are priorities to philosophers and, as such, begins with an emphasis on the constructed world of the knower and the relation of that world to "reality." In opposition to a positivist view that claims that there is "an existing truth" and science helps us to get "closer" to it, von Glasersfeld observed:

> ...it is necessary to keep in mind the most fundamental trait of constructivist epistemology, that is, that the world which is constructed is an experiential world that consists of experiences and makes no claim whatsoever about 'truth' in the sense of correspondence with an ontological reality. (1984, p. 29)

According to this view, "reality" for the individual is constructed based on the individual's experiences; knowledge is evaluated based on its "fit" with those experiences. There is no external, true reality constructed in exactly the same way by every individual. From this starting proposition, constructivists then need to generate explanations for how it is that individuals can come to a shared understanding of reality (e.g., Cobb & Yackel, 1996; Steffe & Kieren, 1994; von Glasersfeld, 1991).

Ultimately, both perspectives, the positivist and the postmodernist, assume the existence of an ontological gap between the object and the subject. From an epistemological perspective, this gap implies fundamental differences, for example, between radical constructivists, who believe knowledge is first constructed internally and then externalized; and socioculturalists, who believe knowledge is first external and then internalized. Yet, the implications that can be derived from either claim simply have not provided useful information for the main decision-making issues confronted by teachers, curriculum developers, teacher educators, and researchers. In an attempt to give more useful information to researchers, teachers, and others, socioconstructivists agree that these theories can be used simultaneously, even though they are contradictory at an epistemological level (Cobb, 1994); however, the dilemma of the difference between the perspectives is not resolved (Kieran, 2000; Lerman, 1996, 2000; Steffe & Thompson, 2000).

Other fields have also been preoccupied with these ontological questions and ultimately arrive at different practical outcomes. Latour (1990, 1999) claimed that the realist's and the relativist's assumption of the ontological gap that separates language from the world is erroneous. Latour asserted that there is no such gap and that what has been regarded as an ontological gap that needs to be leaped to get from the world of experience to the world of representation turns out to be a series of translation processes involving entities in the world and the words used to describe those entities. The difficulty that Latour found with realist and relativist perspectives is pragmatic.

Despite the complexity that can arise, a models and modeling perspective begins with simple, straightforward, and practical assumptions that are further described in the following paragraphs: (a) people interpret their experiences using models; (b) these models consist of conceptual systems that are expressed using a variety of interacting media (concrete materials, written symbols, spoken language) for constructing, describing, explaining, manipulating, predicting or controlling systems that occur in the world; and (c) models developed in and for the world are constantly interpreted and reinterpreted.

According to this models and modeling perspective, the smallest unit of epistemological analysis is a model. A conceptual model is developed by an individual to construct, describe, or explain their mathematical experiences. It can be thought of as having both internal and external components. What we ultimately observe are the external components (representations), but these cannot be disengaged from the internal conceptual systems. Our interpretations result from the interactions between our models and the systems they describe. Every model has some characteristic that the described system does not have; and, every model does not have some characteristic that the described system does have—otherwise the

TABLE 1
The Nature of Reality

Constructivism	*Models and Modeling*
In mathematics education, a number of the most significant leaders in the constructivist movement emphasize that constructivism is primarily a theory designed to address issues that are priorities to philosophers. Constructivism begins with an emphasis on the constructed world of the knower and the relation of that world to "reality."	Rather than beginning with the claim that *the reality that you think you see isn't really out there*, a models and modeling perspective begins with more simple and straightforward assumptions. That is: (a) people interpret their experiences using models; and (b) these models consist of conceptual systems that are expressed using a variety of interacting media (concrete materials, written symbols, spoken language) for constructing, describing, explaining, manipulating, predicting or controlling systems that occur in the world.

model and the described system would be the same. Multiple media are often required to represent complex systems. The development of representations of conceptual systems often causes reinterpretations of the described system as well as the model itself.

There are often many layers of interpretation between an individual's interpretation of an experience and the experience itself. These layers of complexity are a result of the multiple models that may be used to interpret an experience. Most of what we understand about complex systems derives as much from the models developed to explain them as from the existing system itself. Our interpretation is what we know about complex systems; and, our models are used to interpret and to reinterpret such systems (refer Table 1).

THE NATURE OF KNOWLEDGE

Since the work of Piaget focused on describing the nature of the constructs (or conceptual systems or schemas for making sense of experience) that underlie children's knowledge and reasoning processes, it was inevitable that construction emerged as an important process in development. The constructivist claim that all knowledge must be constructed by the individual tends to further emphasize the importance of construction. This has led to one of the most typical generalizations associated with constructivism, namely the notion that all knowledge must be constructed. We wish to make two distinct objections to this claim: (a) there exists important knowledge that is not in the form of constructs and (b) construction is only one of many relevant processes in knowing.

Significant portions of the knowledge that students need to learn are not constructs at all. Some knowledge (such as skills and procedures) is not constructed in any meaningful sense of that term. Skills and procedures are not constructs in the sense of conceptual systems and hence they do not need to be constructed (e.g., Dark, 2003; Lesh & Doerr, 2000). Children certainly need to acquire proficiency with skills and procedures, some of which will be developed into larger more complex skills and procedures. These may be later used in the construction (or reorganizing or testing or refining) of constructs (or conceptual systems), but the skills and procedures themselves are not constructs. The complexity of a skill (just like the complexity of the technical skills of a surgeon or a violinist or a financial analyst) should not be confused with or mistaken for the conceptual understanding of a system with its objects, operations, and relations.

Important knowledge is gained in many ways that are not "constructed." Piaget himself emphasized that constructs are sorted out and reorganized or reconceptualized at least as much as they are assembled (i.e., constructed) by smaller constructs. Constructing is far too narrow to describe the many ways and nuances of ways that significant conceptual systems are learned. For example, sorting and selecting among competing ideas, refining or revising ideas that conflict with other perceptions, filtering out signal from noise (when you do not even know that there might be a signal to look for yet!), selecting pieces of information to work on, representing relationships that are already understood in some partial sense, and so forth are much more than the construction or assembly of previously learned schemas. In other work, we have described the learning processes beyond construction that children engage in when confronted with the need to create significant mathematical systems (models) to describe or explain experienced phenomena (English & Lesh, 2003; Harel & Lesh, 2003; Zawojewski & Lesh, 2003). We have seen numerous examples of students filtering out irrelevant information, differentiating among conditions in which a particular solution would be applicable, integrating previously considered quantities or relationships into current systems of interpretation, and restructuring representational media in ways that would be useful to other people. From a modeling perspective, these are the relevant processes, of which construction is only one, that are involved in the development of a model or conceptual system.

"Construction" does not capture many of the most important activities that students need to engage in when learning mathematically significant conceptual systems. For example, complex notational systems (such as the Cartesian system) and symbol systems (such as the language of functions, limits, and continuity) do not in themselves need to be constructed by each individual (although certain meanings associated with them may well need to be constructed). Engaging students in the "discovery" or "invention" of such systems is likely to miss the central point of learning these systems. Bruner (1960) emphasized that conceptual development often had to do with constructs and conceptual systems that are understood using one type of rep-

TABLE 2
The Nature of Knowledge

Constructivism	*Models and Modeling*
A common statement of the basic principles of constructivism is the claim that "knowledge is actively created or invented by the child, not passively received from the environment." (Clements & Battista, 1990, p. 34)	A modeling perspective generally agrees with constructivism that constructs must be developed by students themselves, but it's misleading to suggest that all constructs must be constructed. Construction is only one of the relevant processes. Constructs are sorted out and reorganized at least as much as they are constructed. Significant portions of the knowledge that students need to learn are not constructs at all and do not need to be constructed.

resentation system (enactive, iconic) gradually being reconceptualized using a variety of written or spoken symbolic representations. If such representational systems are given or "told" to students, the central activity that students need to engage in is the unpacking of the meaning of the system and the flexible use of the system in ways that enable them to make sense of their experiences. For example, in understanding the Cartesian coordinate system or other graphical representations of data, students need to understand when the particular representation is useful, when it is not, what it represents, what it misrepresents, and how to meaningfully interpret their own representations and those of others. Similarly, when students are given a tool such as a spreadsheet or graphing calculator, the task for the teacher and the student is to understand the operations and representations that are possible with the tool, to use these operations and representations in ways that are appropriate to given contexts, and to support their own mathematizing of situations in ways that are powerful and generalizable.

In summary, from a models and modeling perspective not all knowledge is in the form of constructs that must be constructed by each individual and construction is only one of many processes that is used to develop knowledge. Since constructivism and a models and modeling perspective have different views about the nature of knowledge, there are also differing ideas about how knowledge is developed (see Table 2).

THE NATURE OF DEVELOPMENT

Piaget was one of the earliest researchers to reveal the nature of some of the most important constructs that students must develop to make sense of quantities such as simple counts, length and area, speed, force, and probabilities. Over the past 30 years, much constructivist research in mathematics education has generated careful studies of the nature of the development of students' understanding of these constructs. Researchers have provided detailed descriptions about the sequences and stages that students go through in content areas such as whole number arithmetic (Carpenter, Franke, Jacobs, Fennema, & Empson, 1998; Fuson, 1986, 1990; Steffe & Cobb, 1988), rational number and proportional reasoning (Lesh, Post, & Behr, 1989), and functions (Confrey & Smith, 1995; Tall, 1992; Vinner & Dreyfus, 1989). This work has, in turn, been followed by the development of curricular materials and teaching approaches that guide students' constructions along research-based learning trajectories (e.g. *Cognitive Tutor*™ by Carnegie Learning; *Connected Mathematics*, 1991; *Mathematics in Context*, 1998; and others).

The construction metaphor suggests that constructs are assembled in the minds of learners and the task of the teacher is to provide a set of experiences to guide this assembly along a path by beginning with what it is that the child already knows. Such an understanding of construction borrows from the mechanistic and machine-based metaphors of the industrial revolution and the information-processing era. These metaphors suggest that the construction of knowledge in a child's mind is similar to the process of assembling the parts and subassemblies of a machine or programming a computer. Evidence of such metaphors can be seen in approaches to the teaching and learning of mathematics that present children with carefully delineated sequences of instructional materials that are intended to guide the children's thinking through some sets of experiences (often with concrete materials for younger children) toward an adult's way of thinking about the construct (Fennema & Carpenter, 1996; Franke & Kazemi, 2001; Yackel & Cobb, 1996). Carefully guided questioning techniques are intended to move the learners' thinking toward the teacher's or researcher's way of thinking about the construct or problem. A hallmark of such approaches is the notion that development of ideas is sequential and linear, like a ladder, or along a trajectory planned by researchers or teachers.

A models and modeling perspective has a different view of the nature of development (see Table 3). In important mathematical areas such as ratios, rates, proportional reasoning, and other fundamental concepts involving measurement, geometry, algebra, or calculus, it has become increasingly clear that: (a) primitive understandings of relevant models often begin quite early (e.g., at the same time that many whole number arithmetic concepts are at early stages of development), (b) the development of any given model generally continues over time periods of many years (rather than applying to all contexts over a short period of time), and (c) at any given point in time, the evolution of a given model tends to involve a number of dimensions (e.g., concrete–abstract, simple–complex, intuitive–formal, situated–decontextualized). Consequently, a child whose performance on one task appears to be at stage N to a constructivist frequently appears to be at a completely different stage for a slightly different task; and, even within a single learning or

TABLE 3
The Nature of Development

Constructivism	*Models and Modeling*
Studies focus on careful descriptions of how learners understand a particular construct at a particular point in time.	Studies are more focused on knowledge development along multiple dimensions, often simultaneously and in no particular order.
Many researchers have done careful and elaborate studies of the development of students' constructs in areas such as number and ratio.	Students express, test, revise, refine, and extend their own ways of thinking.
Many researchers describe the paths or trajectories along which children's ideas develop.	Development is more about sorting out less useful and less powerful models and less about guiding toward correct models.
Many studies describe various stages of knowledge that children might have or be in and ways to guide them to the next stage.	

problem-solving situation, children often shift from one way of thinking to another without even noticing that they have done so (Lesh & Doerr, 2000). Thus, the apparent stage of reasoning often varies significantly across models, across contexts, and across representations, as well as from moment to moment within a given learning or problem-solving situation. Therefore, it tends to be far too simplistic to describe conceptual development using the metaphor of a single ladder-like sequence or stages.

Instead, conceptual development generally involves multiple dimensions and interactive, cyclic processes. There are a variety of processes involved in development and multiple conceptual systems are developing simultaneously. Various researchers have discussed examples of students thinking along multiple dimensions, using multiple perspectives (often within group settings), and going through cyclic processes of interpretation (Aliprantis & Carmona, 2003; Harel & Lesh, 2003; Lesh, Lester, & Hjalmarson, 2003; Zawojewski, Lesh & English, 2003). Early understandings usually are characterized by fuzzy, fragmented, poorly coordinated, confused, and partly-overlapping constructs that only gradually become sorted out in such as way that similarities and differences become clear. For example, primitive understandings of fractions, ratios, rates, and proportions tend to involve partly formed and uncoordinated conceptual systems that are poorly articulated (using spoken language or written symbols) as well as situated, piecemeal, and unstable (c.f., Greeno, 1991; diSessa, 1988). In general, it is only at relatively mature levels of understanding that mathematical thinking begins to be organized around abstractions rather than around experiences; and, it is seldom that a given child can be characterized as having only a single concept of (for example) fractions, ratios, rates, or proportions. Instead, multiple relevant models tend to coexist in the thinking of a given child. It is only at relatively late stages of development that these topic areas are sorted out into separate, but linked, conceptual domains.

From a modeling perspective, careful descriptions of how it is that learners understand a particular construct at a particular point in time (while useful for delineating the differences among particular states of understanding) are less useful than descriptions of how knowledge develops across contexts and across time both within an individual and among groups of individuals. Understanding how models change in ways that become better (or more useful) as learners move from simplified understandings that develop into more complexity with more refined and nuanced meanings becomes a central focus of research. Learning is more like the wandering across a complex terrain than following the line of movement along a particular trajectory, although it is certainly true that after the fact the trajectory could be described. The task for researchers is to describe what it means for a model to develop, the dimensions along which that development is occurring, and events (from the perspective of a teacher) that might stimulate or facilitate further development.

THE MECHANISMS OF DEVELOPMENT

In this section we examine the mechanisms that drive the development of knowledge from a theoretical stance of radical constructivism and socioconstructivism, and then show how these ideas can be extended from a models and modeling perspective (see Table 4). In particular, we will describe three types of cognitive conflicts that we find useful in explaining the development of learners' models: (a) model-reality mismatches, (b) within-model mismatches, and (c) between-model mismatches.

TABLE 4
The Mechanisms of Development

Constructivism	*Models and Modeling*
Changes in concepts come about when the learner has to resolve conflicts. Perturbations lead to accommodation and assimilation. Social interactions (related to language and culture) support the development of ideas.	Cognitive conflict includes within-model mismatches as well as model-reality mismatches and between-model mismatches. Models not only need to integrate new information, but often need to be revised in ways that differentiate, rank, and conditionalize existing elements. Communities of ideas are developing—multiple threads develop and interact and operate in parallel.

For Piaget, the modification of the internal structure of concepts was a growth process, evolving from the interactions between internal and external structures. Changes in conceptual systems came about when the learner had to resolve conflicts between perceptions of reality. These conflicts or perturbations in the child's existing way of thinking generated a restructuring or reorganization of existing ideas to accommodate new information. Constructivist philosophers such as von Glasersfeld (1991) expanded the Piagetian notions of conflict to include viability or usefulness. That is, a concept is held by a learner as long as it is viable or useful in making sense of experience, not to the extent that it corresponds with some external reality. Conceptual dissonance (or cognitive conflict) is the primary driving force that promotes the change or development of concepts so that they can continue to be viable or useful in making sense of experience. By rejecting notions of correspondence with external reality, constructivist philosophers focused on the dissonance that could occur within the internal constructs held by an individual.

The philosophical and epistemological difficulty of how it is that constructs and their development go beyond the individual led to the development of social constructivism. In moving beyond an exclusive focus on individual cognition, social constructivists brought in Vygotskian ideas about language and tools and notions of participation, apprenticeship, and communities of practice (cf., Carraher, Carraher, & Schliemann, 1985; Cobb & Bowers, 1999; Confrey, 1995; Lave & Wenger, 1991; Lerman, 1996). This signaled an important shift from individual cognition with its emphasis on the constructs held by an individual to social cognition with its emphasis on acting and knowing as participation in a community of learners. For example, a social constructivist perspective emphasizes that in many problem solving situations, what is problematic for students is both the problem as stated and the social norms that will be used to assess the quality of responses that are given.

Several researchers have focused on how shared knowledge is developed by communities of learners and the influences of this shared knowledge on the cognitive development of individuals (e.g., Bowers, 2000; Cobb & Yackel, 1996; Sfard, 1998; Yackel & Cobb, 1996; and others). The research of Cobb and Yackel pays close attention to how social norms, mathematical norms and sociomathematical norms are established in the classroom. These norms serve to structure the development of mathematical ideas within the discourse of the classroom, as guided by the deliberate actions of the teacher. In part, this resolves the difficulties encountered by the misperception and misapplication of constructivism that "anything goes" in a classroom as long as it is constructed by the children. According to this perspective, the norms of the classroom learning community become the criteria by which constructs are validated. Establishing these norms in ways that are aligned with broadly accepted mathematical thinking is a central role for the teacher. In addition, the teacher's central role is to guide the students' collective thinking along a learning trajectory so as to construct particular mathematical understandings. In practice, this guidance is often in the form of carefully posed questions by the teacher accompanied by the selection of responses that further the movement along the path that the teacher has chosen toward a convergence of meanings that become taken-as-shared.

From a models and modeling perspective, we want to move the focus away from the guiding role of the teacher in creating shared meaning in the classroom and to focus on the (multiple) roles of the students in developing multiple threads of ideas which develop, interact and operate in parallel. We have seen this repeatedly happen with the use of modeling tasks that begin with the creation of diverse student generated ideas. With the use of these activities, the role of the teacher is not to select among competing ideas so as to lead the students to shared meaning, but rather to design the learning environment in such a way that the students themselves can test, sort, select, and revise among competing ideas. Representations (such as symbol systems) take on meanings as individuals create symbols, act with them, and interpret the symbols to each other and to themselves. As we have seen in Zawojewski, Lesh, & English (2003) on the roles of group members in modeling situations, either a group can provide multiple perspectives or a given individual can take more than one perspective on a problem situation. The point of modeling is to engage students in the process of selecting among alternative ideas, varying representations and differing perspectives so that the model is useful (a) for a purpose that is understood by the learner and (b) by others who may not have been involved in the creation of the model in the first place. A model is not simply the construct of an individual or for an individual. Useful models can be used by the individual or group who created them in the first place, but they can also be used by others.

From a models and modeling perspective, we find it useful to identify three distinct types of cognitive conflict: (a) model-reality mismatches (such as those that occur when predictions do not match reality); (b) within-model mismatches (such as those which occur when attention shifts from one aspect of a representational media to another); and (c) between-model mismatches (such as those which occur when two distinct ways of thinking about a problem do not agree). From a philosophical perspective, we readily agree that model-reality mismatches implicitly involve within-model mismatches and that the distinction between within-model mismatches and between model-mismatches is purely in the eye of the beholder. However, when working with teachers, curriculum developers, researchers and others who are involved in the development of significant mathematical models by children, we find it useful to make the above three distinctions.

Mismatches with reality from any of these perspectives may lead to changes in the learner's model. It has been well documented, for example, that learners' naïve theories of motion are similar to Aristotelian ideas of motion and do not agree with Newtonian models of motion. However, these naïve theories of motion do serve the purpose of explaining and describing aspects of everyday perceptions of the motion of objects. They do not serve well, of course, the purposes that are better served by Newtonian laws of mechanics. Hence, the model-reality mismatches that lead to changes or improvements in a model are not necessarily the mismatch between a learners' model and that of an expert. Rather, they are the mismatches that occur when the model is less than useful for the learner's purposes in describing or explaining experiences.

Thus, an important element in the notion of model-reality mismatch is the purpose for developing the model. Asking "if it is useful" or "how well does it fit" are simultaneously questions from the perspective of (a) the learner who developed the model, (b) someone else who might use the model, and (c) someone using the model in a related situation. Many modeling tasks have as part of their purpose the design of a system that can be useful to someone other than the person who created the system. In this sense, the usefulness or fit with reality extends beyond the particular situation at hand for the learner to a set of perspectives that may not have been immediately obvious to the learner at the beginning of the task.

Mismatches can also occur within a model. These within-model mismatches are especially likely to occur when learners explore their representational systems and when technological tools are used in modeling tasks. Bruner described this exploration of looking at what you have just done and then representing it to yourself, when he asks: "How can I know what I think until I represent what I do" (1960, p. 101)? We have illustrated this knowing through representing to one's self in earlier research with middle school students on the Summer Jobs problem (Lesh & Doerr, 2000). In this modeling task, we saw how students revised their notions of a "best worker" when they saw the restructured representations of their data in a spreadsheet. The interpretation of new quantities generated by the use of representational tools led to changes within the students' model and to its further development in ways that were increasingly differentiated.

Within-model mismatches also occur as learners engage in the process of integrating new elements into an existing system, reorganizing a system to better account for objects and their relationships, and differentiating among conditions, meanings and consequences. These activities become the source of within-model mismatches that drive forward the continued development of the model. We found that students' first solutions to the O'Hare Airport modeling task became increasingly differentiated as the students needed to account for new elements reflecting the various conditions and constraints that would be used within their model (Lesh, Cramer, Doerr, Post, & Zawojewski, 2003). Modeling tasks require students to fit the elements of the model with each other by expressing relationships, quantifying operations, and constructing new quantities. The integration of elements within a model, the resolution of discrepancies among represented elements, and the interpretation of new quantities are the mechanisms that drive forward the development of the learner's model in ways that are continually more useful to the learner, but without some predetermined path of development as the guiding framework for the child's work.

Between-model mismatches occur when one model is contrasted with another. Usually, when the individuals become aware that two models are inconsistent, they begin to notice some of the differences, and either determine which of the models appears to be "fittest" (for the specific purpose and context), or develop a new model that incorporates elements from both of the models as well as entirely new characteristics. An example of this can be seen in a transcript of the Paper Airplane Problem (Zawojewski & Carmona, 2001), when three students working on the same team are defining what it means to be a "floater" paper airplane. One of the students gives an initial definition, which is debated and refined by a second student on the team, and this second definition is again debated by the third student, and refined to a definition that incorporates elements given by all of the members in the team.

THE ROLE OF CONTEXT AND GENERALIZATION

Claims about the situated nature of knowledge have emphasized that knowledge is organized around experience at least as much as it is organized around abstractions. Situated theorists (e.g., Brown, Collins, & Duguid, 1989; Greeno, 1991; Lave & Wenger, 1991) argued that the nature of the knowledge that learners develop is strongly influenced by the social situation in which the learning takes place and by the particulars of the learning activity. As Putnam and Borko (2000) pointed out, "How a person learns a particular set of knowledge and skills, and the situation in which a person learns, become a fundamental part of what is learned" (p. 4). Much of the research on the situated nature of knowledge has examined the extent to which skills and procedures (rather than conceptual systems) are deeply situated in the context in which they are learned. From the situative perspective, knowledge is developed and exists within a context. The context organizes the knowledge. It has been well-established that situated skills and procedures seldom transfer easily to other situations and contexts.

Situated cognition theorists have focused on the role that context plays in the developing knowledge of a learner. From a models and modeling perspective, we acknowledge the critically important role of context in learning, but we wish to shift the emphasis of our focus. That is, we are more concerned with how knowledge is developed and structured to interpret specific contexts. Interpretation of necessity means that you are looking both within and across situations or contexts. Like the situative perspective, a modeling perspective on learning recognizes that learners' models are situated and as such they always preserve (and distort) some

TABLE 5
The Role of Context and Generalization

Constructivism	*Models and Modeling*
Experience and context organize knowledge more so than abstractions and principles—especially for students and novices to a field.	Models are nearly always developed in specific contexts for specific purposes and for specific clients and as such always preserve some elements of the situation.
Even for experts in a field, their conceptual systems often continue to be shaped significantly by the situations that led to their development (Greeno, 1991).	Modeling is structured for the purposes of communication and sharing within and across contexts.

of the elements of the situation in which the modeling activity took place (see Table 5).

In modeling tasks, students are explicitly asked to create generalized systems that can be used beyond the particular situation at hand. In contrast with findings from situated theorists (Brown, Collins, & Duguid, 1989; Lave, 1988; Perkins & Salomon, 1989; Saxe, 1988), it is often the case with model development sequences that the problem facing learners is not one of transfer from one context to another, but that of generalization from a specific situation. Since models are designed to be shared (among peers) and reused (by self as well as others) in other situations, beyond those that led to their development, we find that the problem learners face is less one of transfer from one context to another one, than of generalization from a specific situation.

We wish to distinguish carefully, however, between the claim that the learner has developed a generalized model from the claim that the learner can use the model in a variety of contexts. Students need to learn to discern situations where the model is appropriate and where it is not. During model development sequences, students are provided the kinds of experiences with applying models in new situations that can become part of a repertoire of models and prototypes.

THE NATURE OF PROBLEM SOLVING

Problems in school mathematics are often carefully mathematized and constructed to elicit very specific mathematical relations. The givens and goals of a problem situation are structured so that the irrelevant information has been filtered out. The structure of the problem maps onto a specific mathematical structure. What is problematic for the student in such situations is to find the nonobvious path from the givens to the goals or to recall a lost tool to help solve the problem. Students may be asked for their reasoning about how they solved the problem after their solution is complete.

Problem solving tasks from a models and modeling perspective focus on interpretations of meaningful situations (See Table 6). What is problematic for the student is to develop useful ways to interpret givens, goals, and solution paths. The interpretation process may include filtering and sorting given information and testing and revising possible products. The students' reasoning about the situation is revealed in the process of developing interpretations, explanations, predictions, and descriptions that are related to the problem situation. From a models and modeling perspective: (a) the primary products of problem solving are complex models focused on significant mathematical structures, patterns, and regularities; and (b) the development of such products requires multiple cycles of interpretation.

From our perspective, models (rather than solutions) are the primary products of the learner when engaged in the problem solving involved in modeling tasks. What is central is for the student to develop a model consisting of a conceptual system that is expressed by some representational system and that is useful for some purpose that is understood by the learner. In general, the model represents a process for accomplishing a real world task, which we take to include tasks in the world of mathematics. The model should construct, describe, explain, manipulate, predict or control systems that occur in the world in a meaningful and useful way. The criteria for judging the "correctness" of the model are determined by two main factors: usefulness and generalizability. Models are adopted or rejected because of their usefulness; and useful models (or theories, or conceptual systems) are considered to be those that: (a) begin with assumptions (or "axioms") that are simple and clearly understood, and (b) generate conclusions (or "theorems") that are powerful and not obvious. In particular, the model should be useful to the learner, a user of the model, and a user of the model in a related situation. Usefulness is also determined by the purposes for developing the model. Generalizability is assessed by

TABLE 6
The Nature of Problem Solving

Constructivism	*Models and Modeling*
Problem solving produces solutions that are often one-word, one-number, or one-sentence answers.	Problem solving produces models or complex tools.
Students' reasoning about the solution is usually asked for after the problem has been solved.	Students' reasoning is revealed in the process of interpreting the problem situation and creating descriptions, explanations, and predictions.
Students construct the product along (sometimes predetermined) ladder-like, linear, sequential stages.	A model is assessed by its usefulness and generalizability.
	Models are developed in multiple interactive cycles similar to other types of knowledge development.

determining how useful the developed model is for the client in other circumstances that differ from the original given situation.

The models students develop during modeling tasks are much more complex and powerful than one-word or one-number answers, and hence the development processes are more complex than ladder-like, linear stages leading from givens to goals. Because the goal of problem solving is to develop a model incorporating a conceptual system, the descriptions of cyclic, interactive problem solving processes are very similar to descriptions of knowledge development. The mechanisms that aid knowledge development also aid problem solving since students are developing knowledge while they are solving the problem.

Developing a model requires multiple cycles of interpretation. Each shift from one way of thinking to another represents movement from one cycle to another. As with early stages of knowledge development, the initial understanding of a problem is often fragmented, poorly coordinated, piecemeal and confused. The evolution of the model developed occurs along multiple dimensions (e.g. concrete–abstract, intuitive–formal, and so on). The same types of model-reality, within-model, and between-model mismatches that encourage conceptual development help students revise and test the models they develop to solve problems. The group environment for problem solving provides the opportunity for students to communicate their models to the other students and thereby provides a setting in which between model mismatches are likely to occur (Zawojewski, Lesh, & English, 2003). After going through multiple cycles in the problem solving process and resolving model mismatches, the finished product represents a more complete, complex solution than the students' first way of thinking about the problem.

THE NATURE OF TEACHING

Numerous researchers have observed that constructivist theories of learning are not theories of teaching (Davis, Maher, & Noddings, 1990; Simon, 1995). At the extreme, one could argue that from a constructivist viewpoint, all teaching is constructivist, since students must necessarily construct their own knowledge. However, such a point of view is hardly useful for supporting the meaningful decisions of teachers or providing principles for the design of instructional materials.

From a models and modeling perspective, we argue that the model development principles that apply to students' learning also apply to teachers' learning about students and about teaching (see Table 7). Thus, our perspective has direct implications for the role of teachers in implementing and designing learning environments, for the development of curriculum materials that can help teachers see and interpret the complexities of teaching and learning, and for the design of teachers' professional development programs.

TABLE 7
The Nature of Teaching

Constructivism	*Models and Modeling*
Knowledge cannot be transmitted.	The same principles apply to teachers as to children in their learning.
The teacher's role is to guide children's thinking along ways that are understood as mathematically valid by the teacher.	The teacher's role is essentially to nurture the conditions in which children's model development can occur.
	Teachers' constructs are more complex than children's constructs and we know very little about the constructs of teachers.
	Teaching is as much about "seeing" and interpreting situations as it is about "doing."

The long history of process–product research has attempted to directly link specific teacher actions with particular student outcomes. The search for such mechanisms reflects a view of the teaching and learning processes that are akin to the factory of the early twentieth century. Such a view of teaching focuses preservice teacher education and continued professional development on the pragmatics of skills, procedures, guidelines and "rules of thumb" that can be used in the classroom. However, focusing on what it is that teachers do seldom provides insight into how it is that teachers think or even what aspects of the teaching and learning process that teachers are thinking about. Fortunately, the mechanistic, assembly metaphors in the realm of teaching practice are gradually being replaced by more ecological, systems metaphors where the development of teachers' knowledge is seen more like the experience and growth of a complex, biological community. This latter sense of the development of teachers' knowledge is at the core of the models and modeling perspective put forward in this article.

Teaching mathematics is much more about seeing and interpreting the tasks of teaching than it is about doing them. Indeed, a distinguishing characteristic of excellent teaching is reflected in the richness of the ways in which the teacher sees and interprets her practice and not just in the actions that she takes. It is precisely a teacher's interpretations of a situation that influences when and why as well as what it is that the teacher does. The essence of the development of teachers' knowledge, therefore, is in the creation and continued refinement of sophisticated models or ways of interpreting the situations of teaching, learning and problem solving. We believe that the theoretical constructs that govern the development of useful models by students are the same theoretical constructs that govern the development of useful models by teachers (Doerr & Lesh, 2003).

The knowledge development referred to earlier in this article can be seen as the learning of teachers or the learning of students. Like students, teachers need to develop systems of interpretation that account for their experiences. For teachers, this includes seeing children's thinking, responding to that thinking in ways that further it along multiple dimensions, differentiating the nuances of particular learning contexts and materials, developing principles and more generalized understandings that cut across teaching contexts and situations, and revising their own interpretations in light of evidence from experience. Understanding teachers' thinking as their models (or systems of interpretation) of experience means that these models need to be understood (just as children's models can be understood) as developing along multiple dimensions, as driven by mismatches with reality and within the model, and as being shared and reused within a community of practitioners.

This perspective has powerful implications for how we understand the role of the teacher and for how we support the professional development of teachers (Koellner Clarke & Lesh, 2003; Schorr & Lesh, 2003). The teacher's knowledge needs to include an understanding of the multiplicity of children's models as those models develop along multiple dimensions. The teacher's role is not to guide students through careful questioning along preplanned learning trajectories, even when these trajectories are revised as the lessons unfold. Rather, the teacher's role is to create the conditions, including the tasks and the tools, that support diverse ways of interpreting problem situations. This allows children's existing concepts to be brought to the fore and new concepts to be developed through processes of selection, refinement, and revision of children's existing ways of thinking. The teacher needs to put students in the situation where they have the opportunity to test their current ways of thinking and see the need to create new ways of thinking about a problem situation. Rather than emphasize guiding the students to move along a path toward an expert's way of thinking about the problem, in modeling situations, the emphasis is on methods of providing the student with opportunities to express his own way of thinking so that those ways can be shared, tested, revised and refined.

Similarly, a modeling perspective on teachers' development focuses on designing effective and sophisticated ways of helping teachers see and interpret children's thinking and support the development of that thinking. Rather than telling teachers what to do or recommending courses of action, we have shifted our work toward ways that will support teachers in their efforts to see and interpret the complexities of teaching and learning. This emphasis on interpretation includes the ways in which students might learn, the mathematics itself, and its pedagogical and logical development, the relevant curricular materials, and possible ways of proceeding with a sequence of learning activities (McClain, 2003; van Reeuwijk & Wijers, 2003). The variability that teachers bring in their perceptions and interpretations of rich contexts for teaching and learning provide diversity in understanding the complex and ill-structured domain of practice. Programs for teachers' development need to provide opportunities for teachers' learning that use this diversity and foster its growth in ways that are increasingly useful to practitioners.

CONCLUSIONS

The models and modeling perspective described in this article suggests several ways of thinking about the teaching and learning of mathematics that moves beyond the practical implications that can be drawn from constructivist theory. Our starting propositions for this perspective are grounded in the notion that people interpret their experiences using models and that these models (or systems of interpretation) are themselves projected back in to the world of experience where they are interpreted and reinterpreted. Like the pragmatic theories of Dewey and others and the constructivist theories of von Glasersfeld and others, for new ideas (or models) to be successful, they needed to be useful or viable in interpreting experience. The models and modeling perspective that we have described attempts to shift focus from questions relevant to educational philosophy to questions that are practical to educational decision makers. A models and modeling perspective seeks to change questions related to the relationship between truth and reality to more pragmatic questions related to the usefulness of a particular model. This implies that the focus of a models and modeling perspective for researchers, teachers, and students is to develop models that explain complex situations in a pragmatic and useful way. Usefulness is judged not only by fit with reality, but also by generalizability, extendability, and fit with other models.

In terms of knowledge development, a models and modeling perspective extends constructivist notions that all knowledge is in the form of constructs and is constructed. Not all of the important knowledge that students need to learn is constructs. Skills and procedures also need to be acquired by learners. Piaget himself recognized that construction was only one of many relevant processes to knowledge development. Knowledge is sorted, refined, modified, integrated, and extended, as well as constructed from existing knowledge. In the early stages of development, knowledge is organized more around experience than abstraction and is far more piecemeal and unstable than Piagetians generally suggest. Moreover, knowledge develops along a variety of dimensions beyond those emphasized by Piaget, who focused on the gradually increasing structural complexity of children's constructs and on development from concrete to abstract understandings. From a models and modeling perspective, we would include the dimensions of situated/decontextualized, specific/general, external/internal, intuitive/formal, and unstable/stable. Children's developing models (as well as teachers' models) are understood as developing aslong multiple dimensions, as driven by mismatches with reality and within the model, and as being shared and reused within a community of practitioners. The role of the teacher, then, is not one of guiding children's thinking along a developmental trajectory, but rather encouraging the generation and expression of a diversity of ideas that learners themselves can test, revise and refine in multiple cycles of interpretation and reinterpretation.

REFERENCES

Aliprantis, C. D., & Carmona, G. (2003). Introduction to an economic problem: A models and modeling perspective. In R. Lesh & H. M. Doerr (Eds.), *Beyond constructivism: A models & modeling perspective on mathematics problem solving, learning & teaching* (pp. 3–33). Hillsdale, NJ: Lawrence Erlbaum Associates, Inc.

Bowers, J. S. (2000). Symbolizing, mathematizing and communicating: A look toward the 21st century. In P. Cobb, E. Yackel, & K. McClain (Eds.), *Symbolizing, communicating and mathematizing: Perspectives on discourse, tools and instructional design* (pp. 385–398). Mahwah, NJ: Lawrence Erlbaum Associates, Inc.

Brown, J. S., Collins, A., & Duguid, P. (1989). Situated cognition and the culture of learning. *Educational Researcher, 18*(1), 32–42.

Bruner, J. (1960). *The process of education.* Cambridge, MA: Harvard University Press.

Carpenter, T., Franke, M., Jacobs, V., Fennema, E., & Empson, S. (1998). A longitudinal study of invention and understanding in children's multidigit addition and subtraction. *Journal for Research in Mathematics Education, 29*(1), 3–20.

Carraher, T. N., Carraher, D. W., & Schliemann, A. D. (1985). Mathematics in the streets and in schools. *British Journal of Developmental Psychology, 3*(1), 21–29.

Clements, D. H., & Battista, M. T. (1990). Constructivist learning and teaching. *Arithmetic Teacher, 38*(1), 34–35.

Cobb, P. (1994). Where is the mind? Constructivist and sociocultural perspectives on mathematical development. *Educational Researcher, 23*(7), 13–20.

Cobb, P., & Bowers, J. S. (1999). Cognitive and situated learning perspectives in theory and practice. *Educational Researcher, 28*(2), 4–15.

Cobb, P., & Yackel, E. (1996). Constructivist, emergent, and sociocultural perspectives in the context of developmental research. *Educational Psychologist, 31*, 175–190.

Confrey, J. (1995). A theory of intellectual development. *For the Learning of Mathematics, 15*(1), 38–48.

Confrey, J., & Smith, E. (1995). Splitting, covariation, and their role in the development of exponential functions. *Journal for Research in Mathematics Education, 26*(1), 66–86.

Connected Mathematics Project. (1991). Retrieved from http://www.mth.msu.edu/cmp

Dark, M. (2003). A models and modeling perspective on skills for the high performance workplace. In R. Lesh & H. M. Doerr (Eds.), *Beyond constructivism: A models & modeling perspective on mathematics problem solving, learning & teaching* (pp. 279–293). Hillsdale, NJ: Lawrence Erlbaum Associates, Inc.

Davis, R. B., Maher, C. A., & Noddings, N. (Eds.). (1990). Constructivist views on the teaching and learning of mathematics. *Journal for Research in Mathematics Education Monograph 4* (pp. 31–47). Reston, VA: National Council for Teachers of Mathematics.

diSessa, A. (1988). Knowledge in pieces. In G. Forman & P. B. Pufall (Eds.), *Constructivism in the computer age* (pp. 49–70). Hillsdale, NJ: Lawrence Erlbaum Associates, Inc.

Doerr, H. M., & Lesh, R. (2003). A modeling perspective on teacher development. In H. M. Doerr & R. Lesh (Eds.), *Beyond constructivism: A models & modeling perspective on mathematics problem solving, learning & teaching* (pp. 125–140). Hillsdale, NJ: Lawrence Erlbaum Associates, Inc.

English, L., & Lesh, R. (2003). Ends-in-view problems. In R. Lesh & H. M. Doerr (Eds.), *Beyond constructivism: A models & modeling perspective on mathematics problem solving, learning & teaching* (pp. 297–316). Hillsdale, NJ: Lawrence Erlbaum Associates, Inc.

Fennema, E., & Carpenter, T. P. (1996). A longitudinal study of learning to use children's thinking in mathematics instruction. *Journal for Research in Mathematics Education, 27*(4), 403–434.

Franke, M. L., & Kazemi, E. (2001). Learning to teach mathematics: Focus on student thinking. *Theory Into Practice, 40*(2), 102–109.

Fuson, K. (1986). Teaching children to subtract by counting up. *Journal for Research in Mathematics Education, 17*(3), 172–189.

Fuson, K. (1990). A forum for researchers. Issues in place-value and multidigit addition and subtraction. *Journal for Research in Mathematics Education, 21*(4), 273–280.

Goldin, G. A. (2001). Representation in mathematical learning and problem solving. In L. English (Ed.), *International handbook of research design in mathematics education* (pp. 517–545). Hillsdale, NJ: Lawrence Erlbaum Associates, Inc.

Greeno, J. (1991). Number sense as situated knowing in a conceptual domain. *Journal for Research in Mathematics Education, 22*(3), 170–218.

Harel, G., & Lesh, R. (2002). Local conceptual development of proof schemes in a cooperative learning setting. In R. Lesh & H. M. Doerr (Eds.), *Beyond constructivism: A models & modeling perspective on mathematics problem solving, learning & teaching* (pp. 359–382). Mahwah, NJ: Lawrence Erlbaum Associates, Inc.

Kelly, A., & Lesh, R. (Eds.). (2000). *The handbook of research design in mathematics and science education*. Hillsdale, NJ: Lawrence Erlbaum Associates, Inc.

Kieran, T. (2000). Dichotomies or binoculurs: Reflections on the papers by Steffe and Thompson and by Lerman. *Journal for Research in Mathematics Education, 31*(2), 228–233.

Koellner Clark, K., & Lesh, R. (2003). A modeling approach to describe teacher knowledge. In R. Lesh & H. M. Doerr (Eds.), *Beyond constructivism: A models & modeling perspective on mathematics problem solving, learning & teaching* (pp. 159–173). Mahwah, NJ: Lawrence Erlbaum Associates, Inc.

Lave, J. (1988). *Cognition in practice*. Cambridge, England: Cambridge University Press.

Lave, J., & Wenger, E. (1991). *Situated learning: Legitimate peripheral participation*. Cambridge, England: Cambridge University Press.

Latour, B. (1990). Drawing things together. In M. Lynch & S. Woolgor (Eds.), *Representation in scientific practice* (pp. 19–68). Cambridge, MA: MIT Press.

Latour, B. (1999). *Pandora's hope: Essays on the reality of science studies*. Cambridge, MA: Harvard University Press.

Lerman, S. (1996). Intersubjectivity in mathematics learning: A challenge to the radical constructivist paradigm? *Journal for Research in Mathematics Education, 27*(2), 133–50.

Lerman, S. (2000). A case of interpretations of social: A response to Steffe and Thompson. *Journal for Research in Mathematics Education, 31*(2), 210–227.

Lesh, R., Cramer, K., Doerr, H. M., Post, T., & Zawojewski, J. (2003). Model development sequences. In H. M. Doerr & R. Lesh (Eds.), *Beyond constructivism: A models & modeling perspective on mathematics problem solving, learning & teaching* (pp. 35–58). Hillsdale, NJ: Lawrence Erlbaum Associates, Inc.

Lesh, R., & Doerr, H. M. (2000). Symbolizing, communicating, and mathematizing: Key components of models and modeling. In P. Cobb, E. Yackel, & K. McClain (Eds.), *Symbolizing and communicating in mathematics classrooms: Perspectives on discourse, tools and instructional design*. Hillsdale, NJ: Lawrence Erlbaum Associates, Inc.

Lesh, R., & Doerr, H. M. (Eds.). (2003a). *Beyond constructivism: A models & modeling perspective on mathematics problem solving, learning & teaching*. Hillsdale, NJ: Lawrence Erlbaum Associates, Inc.

Lesh, R., & Doerr, H. M. (2003b). In what ways does a models and modeling perspective move beyond constructivism? In R. Lesh & H. M. Doerr (Eds.), *Beyond constructivism: A models & modeling perspective on mathematics problem solving, learning & teaching* (pp. 383–403). Hillsdale, NJ: Lawrence Erlbaum Associates, Inc.

Lesh, R., Lester, F., & Hjalmarson, M. (2002). A models and modeling perspective on metacognitive functioning in everyday situations where mathematical constructs need to be developed. In R. Lesh & H. M. Doerr (Eds.), *Beyond constructivism: A models & modeling perspective on mathematics problem solving, learning & teaching*. Hillsdale, NJ: Lawrence Erlbaum Associates, Inc.

Lesh, R., Post, T., & Behr, M. (1989). Proportional reasoning. In J. Hiebert & M. Behr (Eds.), *Number concepts and operations in the middle grades* (pp. 93–118). Reston, VA: National Council of Teachers of Mathematics.

Mathematics in context. (1998). Chicago, IL: Encyclopedia Britannica Educational Corporation.

McClain, K. (2002). Task-analysis cycles as tools for supporting students' mathematical development. In R. Lesh & H. M. Doerr (Eds.), *Beyond constructivism: A models & modeling perspective on mathematics problem solving, learning & teaching* (pp. 175–189). Mahwah, NJ: Lawrence Erlbaum Associates, Inc.

Perkins, D. N., & Salomon, G. (1989). Are cognitive skills context-bound? *Educational Researcher, 18*(1), 16–25.

Putnam, R. T., & Borko, H. (2000). What do new views of knowledge and thinking have to say about research on teacher learning? *Educational Researcher, 29*(1), 4–15.

Saxe, G. (1988). Candy selling and math learning. *Educational Researcher, 16*(6), 14–21.

Schorr, R., & Lesh, R. (2003). A modeling approach for providing teacher development. In R. Lesh & H. M. Doerr (Eds.), *Beyond constructivism: A models & modeling perspective on mathematics problem solving, learning & teaching* (pp. 141–157). Hillsdale, NJ: Lawrence Erlbaum Associates, Inc.

Sfard, A. (1998). On two metaphors for learning and the dangers of choosing just one. *Educational Researcher, 27*(2), 4–13.

Simon, M. A. (1995). Reconstructing mathematics pedagogy from a constructivist perspective. *Journal for Research in Mathematics Education, 26*(2), 114–45.

Steffe, L. P., & Cobb, P. (1988). *Construction of arithmetical meaning and strategies*. New York: Springer-Verlag.

Steffe, L. P., & Kieren, T. E. (1994). Radical constructivism and mathematics education. *Journal for Research in Mathematics Education, 25*(6), 711–33.

Steffe, L. P., & Thompson, P. (2000). Interaction or intersubjectivity? A reply to Lerman. *Journal for Research in Mathematics Education, 31*(2), 191–209.

Steffe, L. P., & Wood, T. (Eds.). (1990). *Transforming children's mathematics education*. Hillsdale, NJ: Lawrence Erlbaum Associates, Inc.

Tall, D. (1992). The transition to advanced mathematical thinking: Functions, limit, infinity and proof. In D. A. Grouws (Ed.), *Handbook of research on mathematics teaching and learning* (pp. 495–511). New York: Macmillan Publishing.

van Reeuwijk, M., & Wijers, M. (2003). Explanations why? The role of explanations in answers of (assessment) problems. In R. Lesh & H. M. Doerr (Eds.), *Beyond constructivism: A models & modeling perspective on mathematics problem solving, learning & teaching* (pp. 191–202). Mahwah, NJ: Lawrence Erlbaum Associates, Inc.

Vinner, S., & Dreyfus, T. (1989). Images and definitions for the concept of function. *Journal for Research in Mathematics Education, 20*(4), 356–366.

Von Glasersfeld, E. (1984). An introduction to radical constructivism. In P. Watzlawick (Ed.), *The invented reality* (pp. 17–40). New York: Norton.

Von Glasersfeld, E. (1991). *Radical constructivism in mathematics education*. Dordrect, The Netherlands: Kluwer Academic.

Yackel, E., & Cobb, P. (1996). Socio-mathematical norms, argumentation, and autonomy in mathematics. *Journal for Research in Mathematics Education, 27*(4), 458–477.

Zawojewski, J., & Carmona, G. (2001) A developmental and social perspective on problem solving strategies. In R. Speiser & C. Walter (Eds.), *Proceedings of the twenty third annual meeting of the North American chapter of the international group for the psychology of mathematics education* (pp. 209–226). Columbus, OH: ERIC Clearinghouse for Science, Mathematics, and Environmental Education.

Zawojewski, J., & Lesh, R. (2003). A models and modeling perspective on problem solving. In R. Lesh & H. M. Doerr (Eds.), *Beyond constructivism: A models & modeling perspective on mathematics problem solving, learning & teaching* (pp. 317–336). Mahwah, NJ: Lawrence Erlbaum Associates, Inc.

Zawojewski, J., Lesh, R., & English, L. (2003). A models and modeling perspective on the role of small group learning activities. In R. Lesh & H. M. Doerr (Eds.), *Beyond constructivism: A models & modeling perspective on mathematics problem solving, learning & teaching* (pp. 337–358). Mahwah, NJ: Lawrence Erlbaum Associates, Inc.